natural connections$_2$

To Jude -
May the magic of nature
fill your dreams!
[illegible]

natural connections$_2$

dreaming of an elfin skimmer

EMILY M. STONE

Green Darner Media
Cable Natural History Museum

Natural Connections: Dreaming of an Elfin Skimmer

ISBN: 978-0-9972061-1-1

Cable Natural History Museum Mission
Connecting people to Northwoods nature through educational experiences that inspire wonder, discovery, and responsibility.

Cable Natural History Museum Vision
We will be a trailhead, community hub, and regional destination with universal appeal. We inspire people toward stewardship of the environment and improve their lives.

Green Darner Media
Green Darner Media is an imprint of the Cable Natural History Museum. Its mission is to connect people to Northwoods nature through publications and other media that inspire wonder, discovery, and responsibility.

To All My Family

Mollie Kreb

Mysterious and little-known organisms live within reach of where you sit. Splendor awaits in minute proportions.

E.O. Wilson

Contents

SUMMER

FALL

Lois Nestel

Introduction

Elfin skimmers are the smallest dragonfly in North America, and one of the many delightful species whom you might encounter in a bog. They are also a spark for my imagination. Deep in my heart, where a little girl still sits under a tree reading fairytales, elfin skimmers dart through the air on shimmering wings. Dragonflies are just as at home in that fanciful land as they are in a scientific treatise. Their chapter became my subtitle because they are a symbol of those things in nature that seem to flow seamlessly between magic and science.

My Aunt Nan was especially good at appreciating the magic in nature. While I was working near her home in Maine (an experience I mention in the chapter "How to Make a Beaver Sparkle"), I brought her on a nature walk and shared my new beaver facts. She hung on every word, and her enthusiasm gave both me and the beaver pond a special sparkle. It wasn't just the beaver pond; she shone her beam of appreciation on everything. Nature is full of many kinds of magical connections. Nan helped me see that, and scientists help me see it, too.

My Aunt Nan always believed that she would come back as a dragonfly, and even from the other side she is one of my biggest fans. In fact, that could be her perched on the cover—beckoning you to appreciate both the magic and the science of nature.

–Emily M. Stone

Sara Qualey

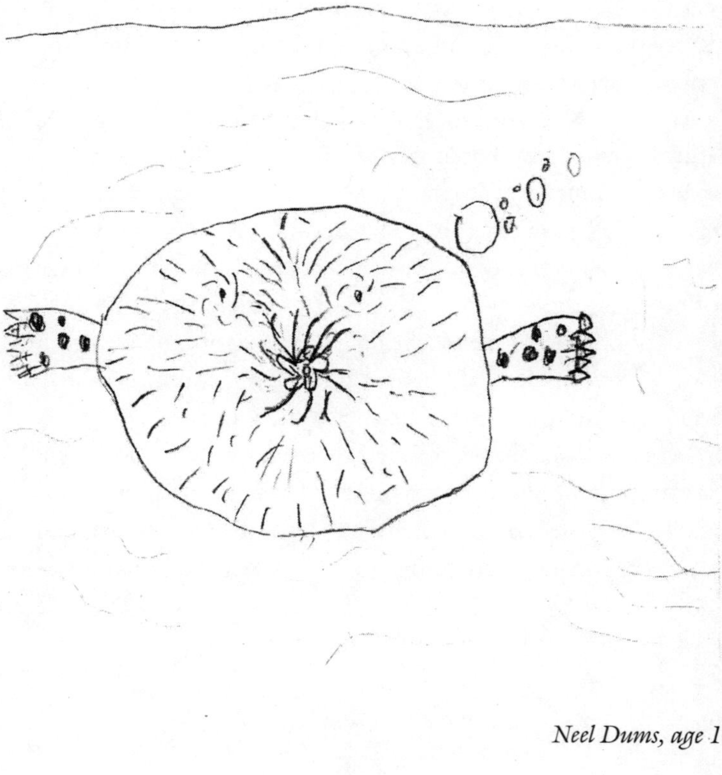

Neel Dums, age 10

WINTER

Hairy Woodpecker in the Fog

Dawn Bronson

Perpetual twilight cloaked the forest on the darkest day of the year. Even at noon, fog hung densely between us and the solstice sun. Cabin fever had set in, and we felt like banging our heads against a wall. Trying to make the best of it, my friend and I layered up and ventured out onto the Superior Hiking Trail. The elegant forms of birch, aspen, oak, and balsam fir near the trail stood out against a plain background. Beyond them, only the gray mist met our eyes. Nothing moved in the oppressive dampness, save us and the assortment of snow, sleet, and raindrops that pelted the world.

Then a bird swooped through the edge of our vision, and a tapping sound punctuated the constant hiss of sleet.

High on the trunk of a dying birch, a black-and-white checkered woodpecker foraged in the bark crevices. His feathers blended perfectly with the white bark and black gnarls. A substantial, chisel-like beak—as long as the profile of his head—was evidently the perfect tool for prying and pecking bark beetle larvae from out of the punky wood. As he leaned back for a better look, or perhaps winding up for a swing at the tree, his stiff tail provided a kickstand for support. The small patch of red feathers on the back of his head was a welcome bit of color in the gray woods, and it also informed me of his proper pronoun.

We didn't stop to watch this male hairy woodpecker, but I did mull over the state of his head as we hiked on. He's obviously quite well adapted for his lifestyle, but how can that little bird bang his head against trees all day and not develop debilitating headaches? Scientists have been studying this problem for years, and for good reason. If we could figure out how to engineer antishock mechanisms so that rapid collisions were less damaging, it would revolutionize the world of transportation safety.

In the case of the woodpecker, scientists in China used CT scans and computer models to discover that the impact energy is converted to strain energy, and 99.7 percent of the strain energy is dissipated throughout the woodpecker's body. The 0.3 percent of the strain energy that affects the head is converted to heat energy. This could cause the woodpecker's brain to overheat and result in a different kind of damage, which is why woodpeckers take frequent breaks while they are pecking.[1] It isn't only about perfecting a rhythm or grabbing a bug, they are protecting their brain.

My psyche could use similar protection from the strain brought on by soupy weather and cabin fever! Instead of taking out our frustration by punching the walls, we hiked faster into the sleet. As the strain dissipated into heat, our red cheeks brought a little more cheer into the fog. It will only get brighter from here.

Woodpecker Niches

Riley Krol, age 7

Just a few minutes up the trail after seeing the hairy woodpecker, we again heard a tapping noise through the fog and sleet. Like déjà vu, a black-and-white checkered woodpecker clung to a birch tree, using its chisel beak and kickstand tail to look for bugs along the knobby twigs. Only this time, I saw no red on its head, the overall size was smaller, and her beak was tiny. Here was a downy woodpecker, a smaller version of the hairy woodpecker we had just seen.

Telling apart these two black-and-white woodpeckers is a rite of passage for beginning birders. My dad initiated me early, with this simple comparison: when you see a hairy woodpecker, you think "What a beak!" because it is so large. When you see a downy woodpecker, on the other hand, you might ask yourself "What beak?" because it is so tiny and inconspicuous. To go with their smaller beaks, downy woodpeckers are only two-thirds the size of a hairy and weigh only one-third as much.

Despite the similarity of their appearances, downies and hairies are not closely related. Scientists think that the resemblance may have evolved because it benefits downies to be mistaken for their bigger cousins. From a distance, the theory goes, hairies have as much trouble telling themselves apart from downies as we humans do, and might choose not to attack when faced with a competition for resources. Up close, hairies have no trouble identifying the smaller downies and chasing them away from bird feeders. Downies still benefit, though, because they can exhibit hairy-sized dominance toward other species of birds at feeders and gain access to more resources.[2]

Their size difference also impacts the places that each woodpecker can forage, and it allows them to exploit almost exactly the same range and habitats while reducing competition.

For example, the feather-light downy (who weighs under an ounce) gleans insects from bark crevices on smaller branches and will often dangle acrobatically on twigs like a chickadee. Meanwhile, the heavier hairy woodpeckers dig for wood-boring insects on trunks and large limbs that won't swing under their weight.

Hairy woodpeckers also specialize by following pileated woodpeckers. After a pileated excavates a large hole and moves on, a hairy might clean up the crumbs of insects that the pileated missed. To satisfy their sweet tooth, hairy woodpeckers have been spotted drilling sap wells into sugarcane and nabbing sap from wells made by yellow-bellied sapsuckers.

Downy woodpeckers have their own set of special feeding habits. Their small bodies and sharp beaks make it easy for them to cling to the dried stalk of a goldenrod plant and drill precisely into a gall on the stem. Inside, they might find a sweet, juicy gall fly larva for dinner. Downies compete with chickadees for this food source. If you find a bird-eaten gall, you can guess at the culprit by the neatness of the hole. Chickadees have much blunter beaks and excavate a messy, wide-angled hole.

When downy woodpeckers forage on trunks they are better able to move horizontally and downward than other woodpeckers. This might give them a different angle on the bark crevices and a better view of food that larger species miss.

Angles are important for the woodpeckers' nest sites too. Both hairy and downy woodpeckers choose a leaning trunk or branch and chisel the nest entrance into the underside. Scientists hypothesize that this keeps flying squirrels or sapsuckers from moving in. Both hairies and downies prefer excavating their nest holes in a living tree with a rotted core or in a soft dead tree, although the downy can use smaller trees and limbs.

The slight but important differences between hairy and downy woodpeckers that allow them to share a forest without pushing each other out are a function of their ecological niches. In shorthand, ecologists sometimes define niche as "how an organism makes a living." More technically, a *niche* can be defined as "the sum of the habitat requirements (including both the living and the nonliving elements of the environment) that allow a species to persist and produce offspring." A niche also includes the ways that an organism responds to the distribution of resources and to competition from other species.

All species have a niche. As my friend and I hiked across a snowmobile trail on the way back to our car, it struck me that the range of human adventurers includes more than just one niche too. With trails for skiing, snowmobiling, hiking, and fat biking winding through the forests, everyone can get out and enjoy the woods in their own favorite way.

Balsam Fir

Ayla Baussan, age 8

The quietness of the woods wrapped around me as I paused for a moment to listen. Under the sunshine and fresh snow, the world gleamed. As my sphere of awareness widened, I realized that the woods weren't as quiet as I'd first thought. My local flock of chickadees chattered and scolded in a birch tree. A woodpecker drummed nearby. And behind it all was the soft patter of twigs releasing their snow load and the plopping of the damp clumps onto the ground.

A clinging, wet snow had blanketed everything. The snow-lined twigs of deciduous trees made artful black and white designs against the sky, but it was a grove of small balsam firs—their pliable boughs drooping gracefully under the icy frosting—that perfected the winter scene.

Most of the year balsam firs aren't highly sought-after. Their wood is soft and brittle. It doesn't hold nails. Pulpwood or light frame construction is about all they are good for.

But this time of year—with snow clinging to everything and holiday carols on the radio—balsams prove their worth. Their conical shape, dark green color, and long-lasting needles make balsams ideal Christmas trees—whether or not they ever make it to your living room. Those characteristics also make them ideally adapted to winter. The graceful droop of their boughs is no accident; that's an adaptation to living in areas with lots of snow. Better to bend than to break. The beautiful green of their needles means they can extend their growing season to the max by being ready to photosynthesize whenever conditions are favorable.

A slight breeze through the treetops initiated a cascade of snow plops onto the trail—and down my neck. That got me hiking again. As I pushed aside a drooping branch, the movement released some of the fir's wonderful perfume. I inhaled deeply and closed my eyes. The smell conjured up memories of a warm cup of tea on a cool morning in the Boundary Waters. Balsam fir tea is a lovely beverage, with potential medicinal value due to the needles' vitamin C content. In fact, balsam may have been used to prevent and cure scurvy, instead of or in addition to northern white cedar, the tree commonly thought of as arborvitae, the "tree of life."

After that fragrance and memory-filled moment, my eyes opened just inches away from the twig. Two layers of needles spread out on either side. Each tiny sprig of green was flattened in cross section. I knew from experience that if I picked off a single needle, I couldn't make it roll between my fingers. Spruce needles, in contrast, are square in cross section, and spinning them between the thumb and forefinger is a nice way to fidget.

"Firs are flat and friendly" was one of the first mnemonics I learned in botany class. As my classmates and I gained confidence with technical terms, the professor added "and they have racing stripes like a fir-rari!" If you turn a fir needle over, two light-colored stripes span its length. These are the stomatal bloom, or coatings of wax surrounding the stomata, which are the pores that allow carbon dioxide in, and oxygen and water vapor out.

I stepped back to admire the tree again. Barely taller than me, it would be the perfect size to decorate for a home. To achieve this height might have taken it nine or ten years. But it could live two hundred more, if all goes well. The tiny seed from which it grew must have found plenty of water in the soil, and just enough light. Balsams are one of the most shade-tolerant trees in the whole forest, and a seedling needs only 10 percent of full sunlight to get started and 50 percent of full sunlight to thrive.

While firs are synonymous with the holidays, they do lack one important decorative item: cones. That's because the seed that produced this fir grew in a cone with deciduous scales; a cone that fell apart. If you are lucky enough to visit a fir grove in early fall, you will see clusters of beautiful, upright purple cones perched near the crowns of mature trees.

Chickadees, nuthatches, squirrels, and porcupines eat the seeds. Moose, deer, and grouse eat the needles. Hungry red squirrels nip buds off the tips of twigs and throw the rest onto my early spring ski tracks. Balsam firs are important wildlife trees all year round.

The trail brought me home, red cheeked and happy. On the front table sat my Charlie Brown Christmas tree—a scraggly little balsam fir that had been struggling in the understory. But when I closed my eyes and inhaled, it too became the "perfect" tree.

A Hike with Auntie Em

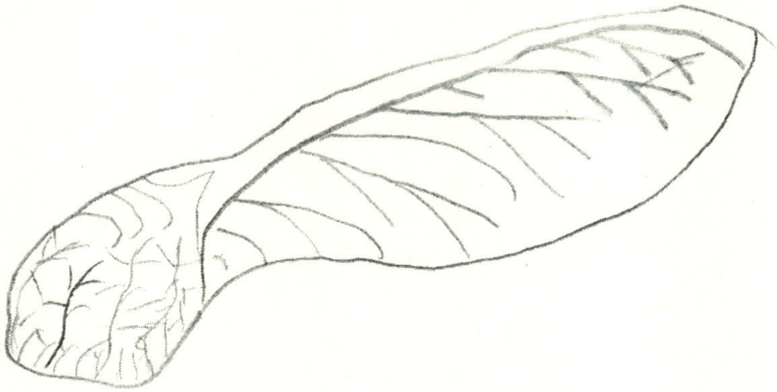

Zac Stone, age 8

"When can we go on a hike with you, Auntie Em?" asked a six-year-old in a shark costume after all the Christmas presents were opened. Those may be the sweetest words I've ever heard. Zac's twin Kylee wanted to come too. After a flurry of finding boots (here's one, where's the other one?), digging grubby old jackets out of the closet, and tugging mittens over small hands, we were off!

With the first big snowstorm for northeastern Iowa still just part of the long-range forecast, we stepped out into a brown world under a blue sky. Zac spotted some pretty flower seed heads and wandered into the restored prairie to pick some for Grandma. "Here's one of those balls," he shouted, grabbing a goldenrod gall from among the flower stems, "and it has a piece of corn in it!"

We examined the gall together. Along the equator of

the small brown globe, a downy woodpecker had used its needle-sharp beak to peck a neat hole and extract the sweet, juicy fly larva. Wedged into that hole was a hulled sunflower seed from Grandpa's bird feeder. This was surely the work of a black-capped chickadee. Those energetic little year-round residents cache as many as 100,000 food items per year—most of them in the winter when food scarcity is a serious risk. To remember all those caches, chickadees add new neurons for every hidden seed, berry, or insect. The result is a 30 percent increase in brain volume, which shrinks again during the easy-living days of summer.

Zac has always had a larger than average head, and I could see it expanding just a little more to accommodate this new bit of information.

Thawing dirt squished under our feet as we turned onto the minimum maintenance road at the end of the driveway. The old road cut deeply into the ridge, with high banks rising on either side. This put cushions of moss at eye level for inquisitive minds. Drawn to the vivid green, Zac got his nose right up into the living carpet. A boy after my own heart. "Helicopter!" announced his first discovery as he grabbed the tiny maple seed.

First he tossed it up, and we marveled at its whirling descent. Then I picked it up and added several feet onto its launch. Zac's eagle eyes followed the seed into the leaf litter, so we had one more launch. This time we passed the seed to Kylee, who had scrambled up to the top of the road cut. Three heads nodded in unison as we tracked the spinning seed. Maple seeds are designed so nicely for human play.

This time the seed landed near a branch; a long, skinny branch with a hooked tip that caught Zac's eye. Both twins worked on getting the gangly tool vertical, and then Kylee backed off and gave orders. "Pull down a tree!" she encouraged, as Zac struggled to hook the stick over low-hanging twigs. Up, up, up he reached, with his every move exaggerated into wide circles at the top. Before long Kylee couldn't stand it, and she joined in to help. With four hands, the hook stayed steady, and finally they got it over a small branch. Who needs plastic toys when you have sticks?

As we detoured off the road onto a deer trail, my pockets began to fill. Zac picked up snail shells, squirrel-sculpted walnut shells, a rodent-chewed chunk of deer bone, and a dozen other trinkets for me to carry back to show Grandma. Young eyes zeroed in on splashes of color in the drab woods. We examined turkey tail fungi coated in bright green algae, discovered a scarlet cup fungus under the maple leaves, and marveled at a rotting stump capped with tiny dots of yellow fairy cup fungi.

Then we found a log populated by puffballs. Once I had demonstrated the effects of poking the deflated brown ping-pong balls, the kids took over. Their small fingers ejected clouds of olive green spores into the breeze. This assistance with spore dispersal was exactly what the fungi were hoping for! Although I don't usually bring up scientific names with kids, I had to tell them that

Lycoperdon, the genus of this mushroom, means "wolf farts." We all giggled before moving on.

The deer trail took us down into a small ravine, where several fallen logs bridged the gap. Kylee, gymnast that she is, headed straight for the first mossy balance beam. I scrambled into the dry creek bed to catch her, but I needn't have. Zac took the more conservative route and scooted across another log on his rear end.

Climbing and jumping off banks, poking at things, and swinging off tree trunks, we made our way back to the house. Zac picked up his bouquet of dried flowers and looked up at me with big brown eyes. "We earned our hot chocolate today, didn't we, Auntie Em?" Yes, Zac, we sure did!

Plenty of Heart

Kylee Stone, age 8

"Piggy back, please!" squealed my niece, Kylee. Because she is adorable, and I don't get to see her often, and she weighs only 38 pounds, I squatted down to let her climb onto my back for about the fiftieth time that day. Only Auntie Em is such a pushover. So with the little monkey on my back, we galloped through the mowed paths of my parents' restored prairie in northeast Iowa. Down to the old barn, past the woodpile, and then back toward the old apple tree.

"Whoa, check that out!" I said as we rounded the side of the tree and noticed that the trunk was only half there. From ground level up to about Kylee's chin, the tree was a partial cylinder of wood with a hollow center. Kylee slid to the ground, and we both stuck our heads inside the cavity to look around.

Blue sky peeked through two holes higher up in the trunk. I circled around to look down in, and by positioning myself at just the right angle, Kylee and I could see each other through the tree, her looking up and me looking down. We giggled.

"Is it still alive?" Kylee asked. "Yes, it is," I replied, "and isn't that amazing?" This old apple tree still produces a bumper crop of the tastiest apples every few years. Since the tree was already here, and already hollow, when my parents bought the land in 1979, we have no idea what variety of apples they are, or when it became hollow.

By the shape of the tree—its straight bottom and leaning crown—I might guess that it was a double-trunked tree that split, or perhaps had a large lower limb that peeled off in its middle age. Once an injury like that opens up, bacteria, fungi, and invertebrates move into the heartwood and slowly decompose it.

All the growth and water transport continues in the outer shell of the tree—the sapwood—even as the center of the tree succumbs to rot. Xylem cells in the sapwood transport water and minerals up from the roots to the rest of the tree. Phloem cells transport sugars from the leaves to the roots and into the growing apples. As new rings of sapwood grow on a tree, the innermost sapwood cells die, fill with resins, and become heartwood.

Losing one's heartwood seems like a devastating prospect, but it actually brings new life to the forest. Tree cavities provide homes for scores of critters. Bluebirds, kestrels, wood ducks, raccoons, flying squirrels, gray squirrels, woodpeckers, owls, wrens, chickadees, bats, mice, porcupines, bears, invertebrates, playful kids, and many more neighbors find shelter in the cozy space of a hollow tree.

For a tree to become hollow, though, it must begin the process while it is still alive. Trees that are not already hollow will not become hollow after they die. The contrast between living sapwood and dead heartwood sets the stage for a great hidey-hole. Cavity trees must get their start early in the life of the forest and have plenty of time to develop. They are a rare and valuable resource.

This hollow old apple tree, with its unique flavors and incredible longevity, is highly regarded among our family. We've grafted its twigs onto two little baby apple trees to try and perpetuate its legacy. We've canned and dried its fruits to feed us through the winter. I'm sure that it's housed a few families of rascally raccoons. And today it inspired wonder and giggles in two little girls. I'd say that it still has plenty of heart.

The Jewels of Summer

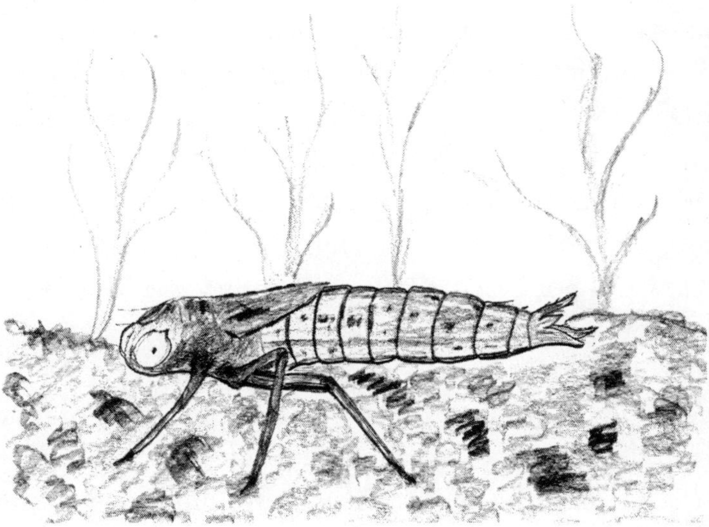

Jane Weber

Bitter winds blow across a frozen landscape, but under the ice hide the jewels of summer.

Even during an Arctic cold snap, many quick-flowing and spring-fed rivers maintain an open channel of inky current. This meandering passage often follows the thalweg (one of my favorite words), which is the deepest channel of the riverbed, usually with the swiftest flow. The input of relatively warm groundwater may help prevent ice formation, as does the constant churning of molecules. Even the still water of lakes and ponds remains liquid below, insulated by the layer of less dense ice floating on top.

Except for the ice fishermen's prey, we often forget that

anything lives in the dark depths below the ice and chilly water. But crawling around on the river bottoms are some of the most grotesque, fearsome, and ancient predators a person may ever encounter. Come summer they will emerge from the depths, shed their gruesome shells, and take flight as shimmering-winged dragonflies.

Understandably, you don't see those colorful acrobats this time of year. Most dragonfly species do not overwinter as adults. The ones that buzzed your hat and caught a ride on your canoe paddle? They're all dead, their genes (hopefully) passed on to the next generation. Common green darners are an impressive exception, as they fly more than a thousand miles to Mexico or Florida with other snowbirds. A few species overwinter as eggs, frozen neatly into the stems of aquatic plants. But most spend the winter as alien-like nymphs hidden under the ice on lakes and streams.

Nymphs are the immature stage of insects that go through gradual or "incomplete" metamorphosis. The alternative is "complete" metamorphosis, which is what butterflies do when they enter the pupal stage and then transform abruptly to a flying adult. Dragonfly nymphs shed their skin several times as they grow through stages called instars until they eventually emerge as a flying adult.

Dragonfly nymphs may spend anywhere from four weeks to several years growing through their instars. The cooler and shorter that the summers are where they live, the longer it takes. In the meantime, they rule the river bottom as fierce predators.

Their hydraulic-powered hunting system is not for the squeamish. To catch food, a nymph draws in water through its anus and clenches its abdominal and thoracic muscles against the water-filled rectal chamber. The amazing amount of pressure now trapped inside the nymph's body cavity pushes out its labium, or toothy lower lip, in a high-speed strike. The lightning attack may earn the tiny predator a meal of tasty mosquito larvae, a tadpole, a small fish, or even another species of dragonfly nymph.

The dragonfly nymph's hydraulic system isn't just used for hunting. By jetting water back out the way it came in, nymphs can propel themselves forward at a speed of four inches per second. That power of acceleration can help when they are on the hunt and also allows for quick exits if they become the hunted. Smallmouth bass, for example, might love a nymph for lunch.

As the water goes in and out, it passes by gills in the dragonfly's rectum and helps the little critter absorb oxygen. This constant filtering of the river through their bodies means that dragonflies can't survive in streams polluted by heavy metals, agricultural runoff, and sewage. Dragonfly nymphs make excellent water quality indicators.

If they survive the winter, dragonfly nymphs will use the abundance of spring and summer to continue growing through their required eight to seventeen instars (depending on the species) before their final mutation into adulthood. The metamorphosis is astounding. From a split down the back of a scraggly, brown bottom feeder emerges a colorful, fairylike being with delicate, dexterous wings.

Their beauty belies the power they retain as a predator. Separate muscles control each wing and enable dragonflies to swoop acrobatically, move straight up or down, fly backward, stop and hover, and make hairpin turns—all at full speed or in slow motion.

Huge eyes take in a 360-degree view and allow them to lock onto a moving target, judge its trajectory, and intercept it—with a 95 percent success rate. Dragonflies' spiky legs, a relic from their previous life as a water monster, form a net to snag prey on the wing. Those prey include our old friends the mosquitoes, who are (for the moment) out of sight and out of mind.

The dragonflies themselves are a bit hidden right now, creeping along river bottoms and lake beds beneath the ice. They are waiting just as we are—lying low until they can tear off their heavy shells and bask in the warmth of the sun. If you saw one, would you recognize it as a diamond in the rough? Could you imagine it soaring as a jewel of summer?

Skiing Through History

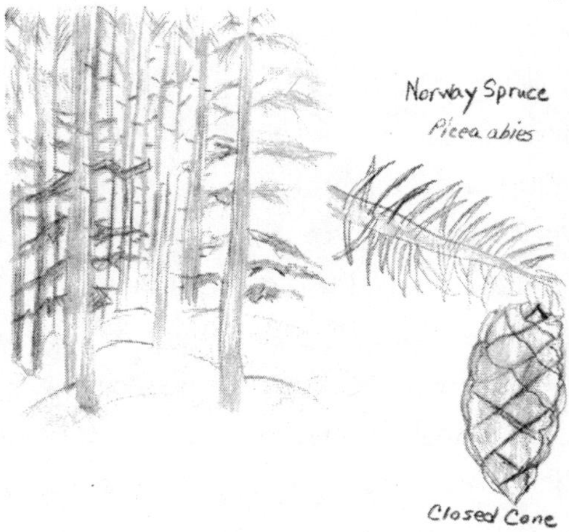

Dawn Bronson

Up hills, around corners, and swooping through the woods I went as my skis swooshed beneath me and poles gripped the trail. A pink sunset glimmered wistfully between the trees, and a pair of ravens gave sonorous croaks as they pushed their wings against chilly air. What a joy to kick and glide over perfect grooming on the North End Trailhead's main 12-kilometer loop.

Earlier, on my first long ski of the season, some parts of the trail felt strangely unfamiliar. Logging operations had altered the scenery by turning patches of dense forest into open woodlands and fields. It was a good reminder that the North End Trails are situated in a working forest owned and managed by Bayfield County.

Just over four kilometers in, not far beyond intersection

76, I paused at the top of a long slope to take a breather. More than a necessary break, this peaceful spot was too pretty to zoom on through. A dense grove of large evergreen trees hugged the trail and seemed to silence the wind.

Looking out, I admired a striking pattern of light and dark stretching as far back as I could see. In their own dense shade, the lower limbs of these trees were dying, and their bare arms held up bright snow. Looking up, I appreciated graceful branches arching up to embrace the sky while their deep-green fingers dangled like the fringe on a shawl.

Those pendulous branchlets are the signature form of Norway spruce, *Picea abies*, and another reminder that this is a working forest. In its native range of Northern, Central, and Eastern Europe (including Norway), this hardy, fast-growing tree is a commercially important source of wood for lumber. Stradivarius used it to make instruments, its cones were once employed as weights in grandfather and cuckoo clocks, and it stands elegantly as the Christmas tree at Rockefeller Center each winter. European immigrants first planted it in Massachusetts in 1860, but its heyday in the Midwest came later.

The "cutover" period, when most of our original forests were logged to build our rapidly growing country, left the land denuded. Hopeful immigrants tried to homestead the land and soon found that the soils were better suited for trees than crops. When farmers couldn't manage to pay their property taxes, their lands were forfeited back to the county. The federal government purchased some land and the county kept some too. On his way out of office in 1933, President Herbert Hoover created both the Nicolet and the Chequamegon National Forests. When President Franklin D. Roosevelt swept in with the New Deal in 1933, a new (and badly needed) era of conservation began.

The Civilian Conservation Corps (CCC) is a well-known product of the New Deal that made an enduring mark on the land. People today still appreciate the log and stone buildings, trails, and bridges that those hardworking boys constructed on all sorts of public lands. Planting trees formed a huge part of their endeavors. A crew of 200 men might plant 150,000 trees per day, five days per week. Their work was essential to reestablishing forests in Wisconsin and most other Midwestern and Northeastern states too.

With the goal of stabilizing the damaged soil of abandoned farms as quickly as possible, foresters experimented with having the CCC boys plant nonnative species in hopes that they wouldn't be susceptible to insect and disease problems that slow growth in native species. Scotch pine, Austrian pine, and Norway spruce were all given a try. When I asked Jason Holmes, a Bayfield County forester, about the success of these foreigners, he admitted that "unforeseen pests weren't very welcoming to these European tree immigrants. Scotch (or Scots) pine was a classic example of this kind of failure." But then he added, "I'm sitting here now wondering if Norway spruce is an exception."

Indeed, it seems to have thrived. When young, Norway

spruce can grow up to three feet per year. It tolerates shade, drought, and acid soils and is not a preferred snack for deer, insects, or mice. Squirrels will nibble on the seeds, but those seeds are still Norway spruce's main route to reproduction, and it has become naturalized in several states.

The calm air I felt while skiing through the Norway spruce grove has been noticed by others as well, and this species is often recommended when planting windbreaks. It's a common street tree too. My neighbor growing up in Iowa had a huge one in her backyard. Once I began looking, I found Norway spruce all over the town of Cable, Wisconsin.

Out on the trail, though, I didn't see much Norway spruce regeneration. The mature, almost 80-year-old trees, which currently represent 40 out of about 170,000 acres of the Bayfield County Forest (a whopping 0.024 percent), will be thinned and harvested according to the same management principles as the native white pines growing nearby. As opposed to aspen, birch, and red maple, Norway spruce doesn't sprout back after it's cut, so these stands should be easy to replant in native species that have returned to favor.

Despite the grove's calm air, I felt the chill start to seep in, and I kicked off down the hill, enjoying the gentle grade of what must be an old logging road. No matter how much I might try to focus on the present—elegant trees, proper technique, and keeping warm—I'll still be skiing through history.

Snow Fleas

Neel Dums, age 10

Late afternoon sunshine flooded through an aspen grove as a group of friends chugged up a hill on our skis. The warmth of a January thaw still hung in the air, and we paused to bask in the glow at the summit. A sprinkling of debris coated the surface of the softening snow. Suddenly one of the little specks vanished. Crouching down for a better look, I discovered that most of the sprinkles were tiny, leaping springtails known as snow fleas.

While springtails have six legs, they aren't insects. Instead, they belong to the insects' sister group, with several different characteristics. A lack of wings, simple instead of compound eyes, differences in molting, a special mouthpart for drinking, and a hinged body part that can fling them into the air set springtails apart from true insects.

A springtail's namesake spring is actually two modified legs on their final body segment, which form a forked appendage called a furcula. While resting, a springtail locks the prongs of its furcula up against its belly, thereby storing potential kinetic energy in its abdominal muscles. A rubberlike protein called resilin in the springtail's exoskeleton helps to flex and store energy there too. The whole system is similar to an upside-down mousetrap.

When a potential predator, such as a giant human skier, startles the springtail, it releases the clasp and the furcula snaps open against the ground. The critter launches into to the air, as far as one hundred times its one-eighth of an inch body length. This erratic movement caused the mysterious, vanishing specks I noticed on the snow. After an uncontrolled flight, a springtail might land mere inches from its starting point, which is still far enough to fool many of its predators—like ants, beetles, and salamanders.

Post fling, springtails don't always land on their feet. Luckily, they have a neat trick for righting themselves. Two inflatable grooming tools not only help them to keep clean but the sticky tubes can also adhere to the ground and be used to pull the springtail upright.

While springtails are in the running to be the most abundant of all macroscopic animals (you might find 100,000 individuals in a square meter of soil, and there are about 3,600 species total), even the most observant people typically notice only the one or two species who spend time on the surface of snow. Snow fleas get their name from their jumpy behavior and the substrate that makes them visible.

On balmy winter days, snow fleas scatter like pepper over the surface of the snowpack, or they might congregate in crawling hordes of a million individuals, drawn together by pheromones for a mating party. It's amazing that such a tiny thing can survive at all in the frozen expanse of snow. The magic that makes this possible is a unique protein especially rich in the amino acid glycine. This protein binds to ice crystals as they start to form, preventing them from growing larger, and so it acts as an antifreeze. The protein works down to about 21 degrees, and at warmer temperatures it disappears! Doctors are studying its potential to help preserve human organs on their way to a transplant surgery.

Snow fleas don't spend much of their life on top of the snow. Like other springtails, they inhabit the soil and leaf litter. Other species of springtails live in tree canopies and even on top of water. A moist environment is essential to prevent desiccation. Since they breathe directly through their skin (another difference from insects, who use a trachea), springtails must balance hydration with breathing. Complex structures on their skin trap tiny, durable reservoirs of air.

In their damp habitats, springtails are essential members of the community. What they lack in size, they make up for in numbers. They fragment organic material and make it easier for decomposers to break down. Springtails also curate the soil's microbial inhabitants by eating bacteria and fungi and by spreading them around. One common fungus—*Laccaria bicolor*—has turned the tables on springtails, though. This fungus paralyzes and decomposes the springtails, then sends their nitrogen over to a white pine through its roots.

On your next sunny adventure in the winter woods, be sure to keep an eye out for tiny, leaping specks on the snow. They might be more interesting than you'd expect!

Grouse House

Sherry Nieds

Two perfect tracks paralleled a sinuous curve of the frozen Montreal River. I was enjoying a leisurely classic ski while my friends all zipped around the wider trails on their skate skis. The morning was perfect: blue skies, bright sun, and temperatures rising quickly out of the teens.

Here and there a few animal tracks stitched their way across the trail. They were coyote, maybe, or red fox. Then a bigger depression beside the trail caught my eye. Was that a sitzmark from a clumsy skier? Not quite. Like an upside-down exclamation point, a yard-long, foot-wide trough ended at a small bridge of unbroken snow. Capping it off was a chicken-sized snow angel.

I slid to a halt and gingerly backed up in the tracks to get a better look. This was the best example of a ruffed grouse's snow roost that I'd ever seen!

Grouse are exquisitely adapted to winter. They grow their own snowshoes in the form of fingernail-like projections on the sides of their toes. Extra feathers on their beaks and legs provide insulation where it's needed. And when their fluffy body feathers aren't enough, grouse use the snow like a cozy down comforter.

Once 10 or more inches of snow accumulate on the ground, grouse will dive or burrow into the fluff and spend the night in an air-filled tunnel. Temperatures under the insulating snowpack often reach a pleasant 32 degrees. No matter how cold it gets, the snow roost rarely drops below 20 degrees. Research has shown that grouse don't need to rev up their metabolism until the temps dip below 25, so the snow roost provides them with significant energy savings.

I could see the grouse's sleeping nook begin where the trough ended. Skiing forward and looking back into the tunnel, I spotted a pile of about a dozen macaroni-shaped scats in the shadows. They represent another purpose for the snow roost. In winter, grouse subsist on tree buds and catkins. Trembling aspen buds seems to be the most nutritious—and grouse prefer them—but the buds and catkins of bigtooth aspen, birch, alder, willow, hazel, and ironwood also provide sugar and protein.

There are two main problems with eating tree buds, though. First, they take a lot work to digest. To address this problem, grouse have a muscular gizzard with added grit that grinds up the woody material. From there, the softer materials enter the caeca. These two extra-long pouches contain specialized bacteria that can digest cellulose and absorb more nutrients before the food exits their digestive system. Waste from the caeca ends up as a dark, shapeless plop, while the woody materials become cylinders of molded sawdust like those I found in the roost.

The second problem is that eating buds in the leafless treetops exposes you to the hungry eyes of predators. Grouse meat is quite tasty. A grouse's solution is to eat an entire day or night's worth of food in about 20 minutes. They store all that food in their crop—an expandable sack at the base of their neck—and then find a safe spot to digest their meal. A snow roost shields grouse from the piercing eyesight of hawks and owls as well as the sharp noses of foxes, coyotes, bobcats, and fishers.

With only my lowly human senses and tendency to daydream, I'm sure the snow roost would have completely concealed its resident from me. If I'd skied by earlier and been the one to flush out this grouse, I would have been astonished by the flying snow and flapping wings.

Snow roosts aren't just an interesting tracking story to read in the snow, though. They are truly a necessity for these birds' ability to survive the winter. Thin or crusty snow—unsuitable for roosts—makes their brown feathers more visible than no snow at all. While a deep freeze may challenge our thickest mittens, it also provides perfect conditions for grouse to hide. Balmy temperatures may feel pleasant on bare skin, but they wreck the snow for grouse and skiers alike. A 2008 study in the *Wilson Journal of Ornithology* found that about 17 percent of the variation in grouse populations is controlled by the snow conditions of the previous winter.[3] Deep, fluffy snow equals more grouse. Crusty snow results in a higher mortality rate.

As I skied on, the tracks of weasels, hares, mice, wolves, deer, and squirrels crisscrossed the groomed trails. Just like the grouse—and the dozens of other skiers out on the trails—they have adapted to winter. If you know how to take advantage of it, good snow can be a real boon to your survival.

An All-Star Nose

Ayla Baussan, age 8

The chickadee took its time choosing a seed. Perched with its tiny toenails pricking the pad of my finger, it cocked its black-capped head this way and that. With its precisely pointed beak, the bird picked up a sunflower seed from my palm, weighed it for a split second, and then tossed it aside. That hull must have been empty.

It tried again. The second seed passed muster, and as the chickadee turned to fly away, its buff-colored belly feathers brushed my thumb. I held my breath—hoping to enjoy the touch of soft down—but my cold, calloused skin barely registered a wisp of sensation. Instead, I got one last poke with spiny toes as the chickadee launched back to the tree.

At times like this I wish I were a mole. Strange, I know, but I don't mean just any mole. I wish I were a star-nosed mole, because they have the best sense of touch of any mammal. They may also be the weirdest looking mammal, and those two things are related. Star-nosed moles get their name from 22 fleshy, pink tentacles that protrude from the tip of their otherwise featureless face, which is on the end of a plain gray, oblong body, flanked by paddle-like digging feet with long, stout claws.

Each tentacle gets its bumpy texture from more than 25,000 touch receptors, called Eimer's organs. Each Eimer's organ contains three types of tactile receptors, two of which also exist in human skin. The third type is unique to the star-nosed mole and allows the mole to identify objects by their microscopic texture. This funny nose may be the most sensitive organ in the entire animal kingdom. If I were a mole, I could have enjoyed even the microscopic softness of those chickadee feathers.

If I were a mole, a vast portion of my brainpower would be devoted to processing tactile information, and I could see only light and dark. I would also eat earthworms. So, maybe the benefits don't outweigh the drawbacks. But star-nosed moles are incredibly well adapted for their own lifestyles, if not mine.

Those supersensing tentacles allow star-nosed moles to touch more than twelve objects per second, creating a tactile map of the environment under their nose. They can "see" without light as they tunnel through moist soil. Sometimes those objects are food, and it takes less than one-fifth of a second (14 times faster than any other mole) for the star-nosed mole to realize that something is edible and then eat it. That is much faster than my seed-weighing chickadee. This skill puts it in the running for the world's fastest eater, right up there with a high school cross-country team at a pasta feed.

Star-nosed moles are even adept at swimming and foraging underwater. This came as a huge surprise to me the first time I encountered this amazing creature. I was snowshoeing on a lake in northern Minnesota on an extremely cold day when I saw something dark on top of the snow. I was confused. Moles live underground and survive the winter by following the worms even deeper underground, right?

Wrong—at least for star-nosed moles. These crazy creatures are active throughout the winter, burrowing through snow and even swimming under the ice of frozen ponds. I can relate. The same winter I found the icy mole, I also dug snow caves and jumped into frozen lakes.

Unlike me, moles use their excellent sense of smell to find prey under the water. Most mammal noses don't work well underwater, because we must inhale air to bring scent molecules in contact with cilia in our nasal passages. To make the life aquatic work, star-nosed moles exhale several bubbles per second onto objects or scent trails they encounter underwater. When the moles draw the bubbles back into their noses, the scent molecules in the air contact olfactory receptors, and voilà! They can smell underwater.

So what good are these little creatures, aside from giving us something to be awed by in the dead of winter? Their tunnels loosen the soil and provide aeration for the roots of plants. Their voracious appetites can help control pest insects, and they provide protein snacks for a wide variety of predators. Owls, weasels, and even largemouth bass ignore the odd appearance of star-nosed moles long enough to gulp them up.

Star-nosed moles may not be as cute as my chickadees, but they can certainly make me appreciate the limits and abilities of my five senses.

Surrender

Roxanne Johnson

You probably know the drill well. First I pulled on the long johns, then the thick, woolly socks, snow pants, a wool sweater, windbreaker, wool hat, ski boots, and—finally—puffy mittens. With a quick look at the thermometer (5 degrees below zero), a short drive to the trailhead, boots in ski bindings, and hands through pole straps, I was off!

Bright sun sparkled merrily over the rolling hills as I shushed along, working hard to glide over the cold, hard snow. In no time, I'd warmed right up, and my core temperature felt more like July than January.

Humans have gotten used to a tropical climate over the millennia, and we try to recreate it wherever we go. Whether it's a woodstove crackling cheerily, our metabolism burning edible fuels, the car heater turned on full blast, layers of wool, down, and fleece trapping warm air around our bodies, or a trip to Florida, we tend to survive the winter by avoiding the cold.

Not everything has that luxury. Clinging to the bark of almost every tree in the forest is an organism that does the opposite. "Lichens master the cold months through the paradox of surrender," writes David George Haskell in his wonderful book *The Forest Unseen*. He compares lichens to the Taoist story of a man who jumps off a waterfall and, to everyone's amazement, returns safely to shore. When asked how he did it, the old man answered, "I go down with the water and up with the water. . . . I survive because I don't struggle."[4]

Lichens don't fight the cold, dry winter air. Unlike wintergreen leaves, they have no waxy coating to slow evaporation. They don't hide roots deep in the unfrozen ground like the oaks. They don't try to fire up their metabolism to produce internal warmth like the chickadees, and they don't grow layers of soft fur to retain it like the ermine.

Instead, lichens allow themselves to gain and lose water as the relative humidity fluctuates. During dry spells, a lichen thallus (leaflike structure) might contain only 15 to 30 percent water, and it goes dormant. Freezing temperatures don't seem to bother them, and at least one species can even photosynthesize at negative 4 degrees. In fact, lichens have survived at least 18 months in the hostile environment of outer space. While orbiting the Earth on the outside of the Space Station, lichens experienced wide temperature fluctuations and the full intensity of UV rays from the Sun. Back on Earth, given water and sunshine, they began photosynthesizing with no residual effects.

On a warm, damp day the lichens I skied past will do the same thing. Their dusty-green, opaque surfaces are filtering out sunlight today to protect inner cells from sunburn. But, with as little as 60 percent relative humidity, moisture will seep back into their cells, the surface will become translucent, and photosynthesis can resume within minutes.

"Plants shrink back from the chill, packing up their cells until spring gradually coaxes them out. Lichen cells are light sleepers. When winter eases for a day, lichens float easily back to life," writes Haskell.[5]

This flexibility to surrender to ambient conditions makes lichens excellent colonizers of harsh environments. On mountaintops, the tundra, fresh rock faces, gravestones, and even mine tailings, lichens are the first signs of life. Vegetation dominated by lichens covers nearly 8 percent of the Earth, largely in the Arctic tundra. What makes them so tenacious?

Lichens are a partnership between two or three different types of organisms. They are a cross-kingdom alliance. A fungus plays host, providing a stable home for a photosynthetic partner. Algae or cyanobacteria (formerly known as blue-green algae) set up shop inside the leaflike structure and use the Sun's energy to make food for them both.

Without the photosynthetic partner, the fungus could not grow in rocky habitats that lack organic matter to decompose. When scientists try to culture the fungus by itself, they end up with a slow-growing lump. Without the fungus, the algae—which need constant moisture to survive and reproduce—would be severely restricted in their habitats.

Together, the fungi and algae in lichens are the quintessential example of a mutualistic symbiosis, where both partners benefit from a close, living relationship. Lichens are also a lesson in Taoist philosophy, exemplifying the concept of *wu wei*, demonstrated by the old man in the waterfall. *Wu wei* means allowing things to flow naturally, without combative or egotistical effort.

A good skier can also use wu wei. Extending the glide, relaxing on the back swing, using momentum, and coordinating movement and breath are all hallmarks of great technique. My more casual technique also includes pausing to enjoy the view (and catch my breath) at the top of a hill, smiling at the bright pink sunset, and finding inspiration in the winter woods.

Winter Houseguests

Jane Weber

The pile of sweaty skiing and outdoor clothes in my closet had begun to smell a little rank, as polypro is inclined to do. So I hoisted the laundry basket and schlepped it down two flights of stairs. In the corner of the last step, something small and dark, about the size of a nickel, caught my eye. "Thunk" went the laundry as I knelt to get a better look. Sure enough, a brown spider crouched among the dust bunnies and dog hairs.

Now I know that this type of discovery elicits a wide variety of responses in people, and the majority lean toward some combo of screaming and smashing. But I'm at the other end of the spectrum. "Seeing" nature has been difficult for me lately, since I spend most daylight hours in the office. As a result, I've been feeling a little deprived of my usual encounters with wildlife, so I felt honored to have some nature living in my stairwell. I ran upstairs, grabbed my camera, and did a little wildlife photography there on the basement steps.

Long, hairy, rusty brown legs were banded with darker stripes. Her oval-shaped abdomen was tan with a squiggly black chevron pattern and more hair. Long, slender pedipalps (also hairy) stood ready on either side of her mouth to taste, smell, and help crush her food. Indeed, it was the delicacy of the pedipalps that identified her as a female—a male would have big, club-like pedipalps to aid in reproduction.

A quick search through Larry Weber's *Spiders of the Northwoods* got me confused on the identity of the little critter.[6] Show me some tree buds and I'll point out minute differences that indicate this twig is (obviously) from a beaked hazel and the other one is from an American hazel. Show me a spider and I say, well, it has long legs and dark stripes. That describes at least a third of the spiders in my guide.

For help, I went straight to the source and emailed the photo to Larry. With enthusiasm, he identified is as "a type of funnel-web spider (funnel weaver); family Agelenidae. She appears to be the genus *Tegenaria*." Then—as with any new roommate—I Googled her to find out more.

According to Wikipedia (a good place to start, but not the final word), "house spiders of the genus *Tegenaria* are fast-running brownish funnel-web weavers that occupy much of the Northern Hemisphere except for Japan and Indonesia."[7] In northwestern Europe, where giant house spiders (*Tegenaria duellica*) are native, the British affectionately call them dust bunny spiders due to their propensity to set up a territory under furniture.

Long ago these spiders might also have chosen to set up territories on ships carrying the first European settlers in the 1600s and in shipments of British lumber during the Napoleonic Wars (early 1800s), thus spreading from Europe to America. Their close association with humans may have earned them passage elsewhere too, and they are now suspected to live in almost every country on Earth.

You have to be fast to colonize the world, and giant house spiders held the *Guinness Book of World Records* for top spider speed (1.18 miles per hour) until 1987. The "spiders" that beat their record are not even true spiders.[8] Humph!

Happily, most of these house spiders would rather try to escape than try to bite you or your pet, so they pose little danger as roommates. David Sedaris, best-selling humorist and author, once befriended a number of *Tegenaria* spiders and noted that they are "as quiet and unobtrusive as Amish farmers."[9] One thing I did worry about was where this spider was living. Oh, I don't care if it sets up shop in the basement stairwell—that wasn't the problem. But everything I read suggested that these funnel-web spiders should be living on a flat, sheetlike web with a tubular retreat at one corner. Nowhere at the bottom of the stairs could I find such a web. So I ventured into the basement itself to see if I could find the spider's true home.

I found the nooks and crannies surprisingly clear of webs until I reached the farthest back corner. There above the phone and internet boxes hung a beautiful gossamer sheet web. Tentatively jiggling it, I was surprised by a tiny black spider that suddenly bungee jumped into the abyss and then lowered himself (male spiders are smaller) smoothly on a line of silk. A second black spider still clung to the web. As I looked even closer, I noticed several pale, leggy cellar spiders clinging upside down to messy, irregular webs like something out of Tim Burton's *The Nightmare Before Christmas*.

The crumpled bodies of dead insects littered the floor under this spider colony. I took this as evidence that these spiders were contributing members of the household, not just some lazy freeloaders. Through all this adventure, my dirty clothes sat abandoned on the floor. If only spiders could add "doing laundry" to their list of helpful talents!

Soft, Gray Wings

Cassidy Judy, age 11

The noiseless glide of soft, gray wings caught my eye. Then, stillness. No matter how hard I squinted, I couldn't resolve the dark shape into a branch and the owl I knew had just landed there. Fading light under gray skies combined with the owl's pattern of pale and dark bars on its chest created perfect camouflage.

Nevertheless, I felt grateful to catch even a glimpse of such a graceful nocturnal creature. On a sunnier day, it might stay hidden against a shady tree trunk until dark, avoiding detection by its nemeses—crows or great horned owls. The gray afternoon and impending breeding season must have lured it from its usual daytime roost.

Most of my encounters with barred owls have been

vocal. Their distinctive eight-note call is often phrased as "Who cooks for you? Who cooks for you-all?" and has earned them the nicknames "Eight Hooter" and "Le Chat-huant du Nord," which is French for "The Hooting Cat of the North." On many childhood nights, conversations between my dad and the barred owls echoed over the Iowa hills. My first attempts at calling in the owls often prompted a response from the local coyote family instead.

As a Girl Scout camp counselor, I remember a very long week in a unit of ten-year-old girls when the barred owls out back decided to practice their full repertoire of calls. When the awful jumble of hoots, screeches, cackles, and yelps broke into a frenzied and raucous monkey-like squall, we ended up with a bevy of scared girls in the counselor tent—and little sleep. I have to admit that the barred owl's voice can still send shivers up my spine. Winter, as their breeding season ramps up, is a good time to listen for their haunting calls.

While their voices can be eerie, the silence of owls is even more intriguing—and lethal. Not only does silent flight enable owls to sneak up on prey, it also allows them to hear a mouse—under a foot of snow—pinpoint its location, and punch sharp talons through the icy crust with deadly accuracy. Any flight noise would obscure the soft scurries of those subnivean snacks.

One secret to silent flight is broad wings with large surface areas that reduce the need to flap. Owls seem to float through the air. In addition, anyone looking at an owl wing can see a comb of stiff feathers along the leading edge and a flexible fringe on the trailing edge of the wing. Those fringes break down air turbulence into smaller currents called microturbulences, which generate less sound.

Looking closely, one may also notice a soft, downy covering on the top of the wing. A photographic study of owl feathers revealed a surprising forest-like geometry of the down material, and scientists at Lehigh University's Department of Mechanical Engineering and Mechanics think that this may be another key to eliminating sound.

Dr. Justin Jaworski, a scientist on the project, predicts that "if the noise-reduction mechanism of the owl down can be established, there may be far-reaching implications to the design of novel, sound-absorbing liners, the use of flexible roughness to affect trailing-edge noise and vibrations for aircraft and wind turbines, and the mitigation of underwater noise from naval vessels."[10]

Just being quiet isn't enough. Owls have especially acute hearing due in part to their satellite dish–like facial disks that funnel sound and to a right ear that is higher on their head than the left ear. The sounds of a scampering mouse—like the ones who have left numerous hopping tracks quilted into the snow near my skis—will reach one ear more quickly than the other ear. By using this triangulation, owls can pinpoint the location of their prey—even below snow, even in complete darkness.

Once prey is located, owls' other weapons kick in. Their sharp talons and hooked beaks are equally strong, unlike other raptors who have stronger feet and weaker beaks. Owls use both to grab, kill, and eat their prey.

The owl was still nearby, I concluded, since I hadn't seen it leave the branch. Although I couldn't see it in the fading light, I knew its huge eyes could be watching me. Owls have excellent binocular vision, which allows them to judge distance and depth. Owl eyes are more than twice as large as the average size for birds of the same weight. In fact, their eyes are elongated tubes held in place—unmoving—by bony structures called sclerotic rings.

To compensate for not being able to move their eyes, owls can turn their heads up to 270 degrees (*not* 360 degrees) left or right from the forward facing position and almost upside down. Owls' 14 neck vertebrae (twice as many as humans have), single pivot point within the cervical vertebrae, special muscle structures, and a special arrangement of the jugular veins all contribute to their flexibility.

Without downy feathers to insulate me, the chilly evening crept into my core. Pushing my skis forward, I tried to catch one more glimpse of the camouflaged critter. This tested the flexibility of my neck (only about 180 degrees), strained my human eyes, and made far more sound than those soft, gray wings.

Goldenrod Gall Moth

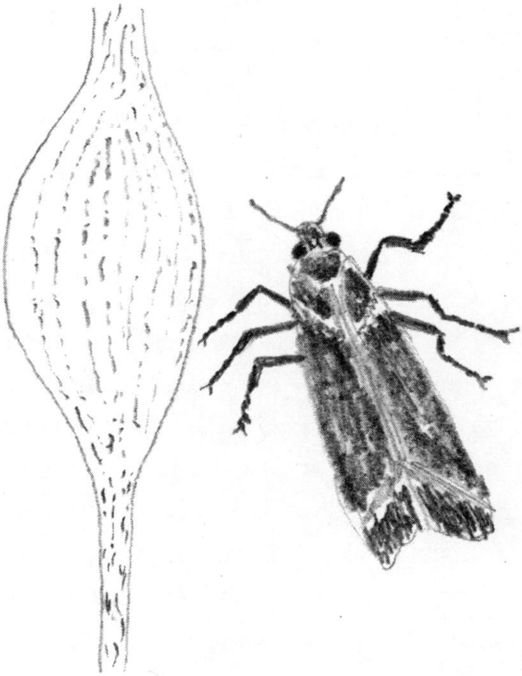

Sue Thurn

This morning—camera in hand—I went looking for an adventure. The otters' early morning tracks went run-run-*sliiiiiide* along the lakeshore. A pair of fox tracks trotted merrily down the driveway on a date (complete with double scent marks!). Shards of snow clung to emerald moss as it clambered over a maple's craggy bark. The morning was beautiful, but still nothing inspired me. So I buckled down at my computer to fill in spread sheets and make graphs.

The data I organized are actually pretty interesting. In

December, sixth graders at Drummond Elementary School in Drummond, Wisconsin, dissected goldenrod galls and recorded their diameter and contents. Most of the time they found the plump goldenrod gall fly larva who created the gall. Occasionally they found a wasp or beetle larva that parasitized the gall larva. I love telling the story, and they love discovering it.

Suddenly, in the middle of copying my final graph into the slideshow for class, I got up with a half-formed thought and wandered into the kitchen. Digging to the back of the freezer I pulled out a folded zip-top bag with a dozen or so brown galls. Here were the leftovers from that December class. These weren't smooth and round like most of the galls we dissected, though. They were oblong and spindle shaped, with some smooth ridges like the bark of a musclewood tree.

Unthinking, I'd handed one out in class with the rest of the galls, and a girl opened it up. Inside the oddly spacious cavity was a large, crispy-dead, dark-brown larva. This wasn't the same insect we'd been studying! I packed the gall and the larva into a petri dish and stuck the rest of the odd galls back into the freezer, intending to research these elliptical galls. But I hadn't gotten around to it until today.

Excited, I took out my trusty dissecting tools: a paper plate and serrated butter knife. One by one I opened the elliptical galls. Empty. Empty. Empty. They were all empty, with large cavities. A white film coated the inner chamber. Did a fungus kill them? Poking deeper, I noticed a small hole near the top of one gall. An exit tunnel!

Finally, the last and largest gall revealed another crispy larva. Under the dissecting microscope I noticed many small holes in the dried exoskeleton. Weird. Now I was hungry for information. Using Google Images, I quickly matched the image of the gall to a name, goldenrod gall moth—*Gnorimoschema gallaesolidaginis*. This frumpy moth has a one-centimeter-long body and two-centimeter wingspan. The forewings are dark, and the hind wings are pale and fringed. The moth is not much to look at, but drama unfolds when you dig deeper. In 1963 William E. Miller of Ohio State University wrote a classic natural history article titled "The Goldenrod Gall Moth *Gnorimoschema gallaesolidaginis* (Riley) and Its Parasites in Ohio."[11] I've summarized his information below.

The goldenrod gall fly larva that my students and I usually study spends its winter frozen inside the gall. This gall moth takes a different approach, and that explains why I found no live larvae in the galls. The adult moth lays its eggs on the lower leaves of goldenrod plants in late summer and fall. The adult dies. The eggs overwinter there. Once a larva hatches in spring, it must seek out a newly sprouted goldenrod nearby. The larva eats its way in through the terminal bud and into the pith of the stem. A few inches below the terminus, it sets up shop. Growth hormone–like chemicals in its saliva trigger the plant to form the elliptical gall.

After feeding inside the gall all summer, the larva chews itself an exit hole, leaving the thin outer skin of the plant intact. Then, prior to leaving, the larva fashions itself a plug for the hole out of sawdust and silk. Brilliantly, the exit hole is wider on the outside and tapers toward the inside. The plug fits like a bathtub's stopper, and it is easy for the moth to eventually pop out but difficult for an invader to push in.

Next, the larva lines the cavity with silk. In doing so, it both seals the chamber and creates a sort of guide to help its future adult self find its way out. The white film I attributed to killer fungi may have been helpful silk! Once both the escape hatch and the runway are in place (sometime in July or August), the larva pupates. The cocoon stage lasts four to six weeks, and then the moth emerges in early fall, pushes open the plug, finds a mate, and continues the cycle.

This ideal scenario rarely happens. About half of the larvae succumb to parasites, mostly parasitic wasps. While reading Miller's graphic description of the various parasites, the meaning of the dried larval skeleton riddled with holes became clear. "*Copidosoma g gelechias* is a polyembryonic internal parasite of the larva," wrote Miller.[12] What does that mean? This tiny wasp lays a single egg inside the moth egg. The moth larva hatches in spring and it grows normally until the wasp egg clones itself and produces lots of wasp eggs. The wasp eggs hatch and are "set free in the host's body cavity."[13] The parasites eat the moth larva from the inside out, leaving only the chitinous hard parts and conveniently waiting to finish it off until after the moth larva excavates the exit hole. Some wasp larvae grow bigger than others, and they attack the larvae of any other wasp that might also have parasitized the moth. In the end, the wasps pupate and emerge, leaving behind a hollow shell riddled with holes—a hollow shell that now sits on my desk.

Driving home past ditches full of dried goldenrod stems, I couldn't stop thinking about this amazing story. How many more stories are hiding in the road ditch?

Martens and Wind

Taylor A. Popelka, age 10

Fierce gusts of wind rocked the treetops above me, sending shivers down their trunks. Among the boles of swaying aspens, maples, and firs, I skied with one ear to the sky. Twigs clattered, snow plopped, and cold wood fibers moaned up and down the musical scale. One tree almost buzzed with a high-pitched popping, while another groaned from deep within. I listened carefully to determine the location of the loudest trees and to gauge their ability to strike me if they came crashing to the ground.

Although the bluster threatened above me, it didn't quite reach all the way down to the swooping hills and rippled snowpack. I felt protected in the understory of this forest. My eyes were scanning for animal tracks in the days-old snow, but I saw only signs of other skiers. Deep drifts and bitter temperatures have driven many creatures down into the next level of protection: the subnivean layer.

Fluffy snow captures summer warmth still radiating from the ground. A thin zone opens up under the snow, right at the surface of the ground. Here, the temperature stays at a fairly stable 32 degrees. Compared to a negative 45 degree windchill at the surface, that feels pretty balmy!

American martens are one of many creatures who exploit the subnivean microclimate. These small weasels tunnel through the snow to find food, stay warm, and escape predators. As a master's student at University of Wisconsin, Madison, Phil Manlick spent the winters of 2013 and 2014 living with me in the Cable Natural History Museum's staff housing and studying the "habitat-mediated foraging and predation risk in reintroduced American Martens." One of his hypotheses was that martens need lots of structural diversity—lots of fallen logs—in their habitat to help them gain access to the subnivean layer.

Once under the snow, martens can forage along the log-lined runways where red-backed voles, mice, shrews, and squirrels also travel. When satiated with a tasty meal, martens have been known to curl up in the den of their prey for a nice, warm nap. The snow is an excellent blanket for this lean mammal who stores little fat and burns lots of fuel to stay warm. Snow also provides cover from other predators. The diets of foxes, fishers, and bobcats overlap with martens' diets, and those larger carnivores will kill martens to eliminate competition.

American martens were extirpated from Wisconsin in the 1920s due to unregulated fur harvest and widespread habitat loss. In 1986 wildlife managers developed a marten recovery plan with the goal of reestablishing two self-sustaining populations of martens in the Chequamegon-Nicolet National Forest. Even with cooperation between the Wisconsin DNR, the US Forest Service, and the Great Lakes Indian Fish and Wildlife Commission (GLIFWC) and multiple reintroductions, marten populations in the Chequamegon National Forest are not rebounding as well as wildlife managers and researchers hoped. Phil tried to figure out why.

One issue limiting the American martens' recovery in the Chequamegon-Nicolet National Forest might be poor juvenile survival. Adults are doing fine, he found, but the young martens, kicked out of their parents' home range after just a few months, may not be finding enough high quality food to make it to adulthood.

Phil's research techniques involved gathering hair samples from martens as they explored PVC tubes baited with beaver meat and equipped with a brush. By comparing the ratios of stable isotopes of nitrogen and carbon in the martens' bodies to the ratios of those isotopes in their various prey species, Phil gained insight into what martens were eating. Other researchers also identified animal remains found in marten scat to record specific prey.

The surprising results are that martens' diets consist of

40 percent shrews and 30 percent deer in northern Wisconsin. In contrast, healthier populations of martens out west and in Minnesota eat many more red-backed voles and squirrels. Why might eating shrews and deer be a problem? Shrews, like the martens themselves, don't store fat for the winter. With an extraordinarily fast metabolism, shrews must feed voraciously night and day. This makes them the skim milk of the small mammals.

Deer might seem like an improbable food source for a two-pound marten, but evidently, martens are feeding opportunistically on winter-killed and wolf-killed carcasses. The trouble is that deer carcasses are a risky food source. Fishers, wolves, coyotes, and other scavengers are attracted to those same banquets and make things dangerous for the martens. That competition with other species may be another important factor limiting the martens' recovery.

Phil suspects that habitat quality has an impact on both food availability and competition. During his summer habitat surveys, Phil found that martens need "nasty forests that you don't want to trudge through." By which he means forests with lots of blowdowns, snags, large trees (greater than 15 inches in diameter), and coarse woody debris. If further data analysis supports his hunch, it might influence how foresters manage the landscape in the future.

Thinking back to those trees groaning and swaying in the tempest above me, it might not be a bad thing if a few of them fell down . . . after I've skied past, of course.

Wolves at Rock Lake

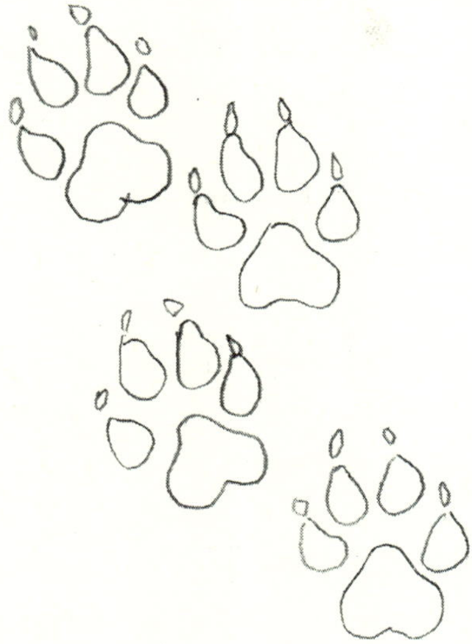

Logan Abric, age 13

The thermometer on my car read 5 degrees below zero when I pulled out of the driveway, but by the time I'd reached the sunny parking lot at the Rock Lake Ski Trails, the temperature was up to zero. A brisk wind hurried through the trees and imparted a sense of urgency to the day—partly because I was rushing to get skiing before the cold seeped in.

Sunshine. Great snow. Roller-coaster hills. Weasel tracks quilting the drifts. I veered right at the first three intersections before finally turning left onto the 11.5 kilometer loop. Lost in thought, I alternated between thinking about my to-do list and thinking about technique. Then shapes in the messy snow along the ski tracks crackled through my subconscious and brought me zooming back to the present. Big feet . . . four naily, untrimmed toes . . . and lots of them. Suddenly I was skiing alongside the tracks of a wolf pack.

They were polite wolves who rarely stepped in the ski tracks. While on the hard-packed trail, they walked freestyle in their own paths. Confined to the narrow strips of smoothly groomed snow on either side of the trail, the mess of tracks on both sides of me indicated that at least five or six wolves had been traveling together.

In a couple locations, the tracks all left the ski trail in single file. Suddenly several wolves looked like one. To save energy, many animals will walk with direct registry. That means back feet land in the tracks of front feet, and members of a line will all step in the same places. Humans do this too, especially in deep snow. Walking in someone else's footprints is much easier than breaking through the crust with each step on your own unique path.

In a few places one wolf did step into the smooth valley of the ski track. His paws were perfectly imprinted there, undisturbed by any other skier. I was the first human to glide over them. These tracks were made last night. I was going their direction. There were wolves at the end of these tracks.

Of course I know that I have nothing to worry about. Wolf packs surround Cable, Wisconsin, and inhabit all the wilds where I play. Tens of thousands of humans recreate in these woods each year, and most don't even see a wolf, much less feel threatened by one.

Instead, the tracks grounded me in the present: in the crisp, sunny woods; in my rosy cheeks; and in the cold air that filled my lungs. When I stopped shushing along to look closer, the sound of wind in the trees filled my ears. Creaking and moaning, snapping and popping, the trees seemed to be composing their own melodramatic poetry against the bluebird sky. The woods felt alive.

When I lived out East, and out West, I missed that sense of thrill that comes from sharing space with the untamed grace of wolves. History says they won't bother me. But biology says they could. I know that not everyone agrees, but I feel incredibly fortunate to live in a place wild enough for them too.

It wasn't always that way. People killed them off once. A small number hid out in the remote areas of northern Minnesota—which holds the only continuous population of wolves in the lower 48 states. They came back to Wisconsin on their own in the mid-1970s. The Endangered Species Act allowed their population to grow and thrive enough that they didn't need that protection anymore. They were delisted in January 2012 and went through three hunting and trapping seasons. In a plot twist, a federal court vacated the 2012 decision in December 2014, which returned wolves in the Great Lakes region to the federal endangered species list. A record high population of at least 925 animals was estimated to inhabit Wisconsin by the winter of 2016–17.

But political controversies don't matter on a sunny day in the woods. Eventually the tracks converged in a narrow trail, loped off through the underbrush, and didn't return. They left a mark on my thoughts though, and when I ran into Sarah Boles, a carnivore tracker for the Wisconsin DNR, the wolf tracks were the first thing I mentioned. After a little follow-up work with the help of our friend Adrian Wydeven, retired DNR wolf ecologist, we determined that I probably saw tracks of the Seeley Hills pack.

The Seeley Hills pack originated in 2003 with wolves that broke away from the Ghost Lake pack. With five to six wolves, including a radio-collared yearling male, 845M, the pack now lives in my backyard. According to Adrian, "The Seeley Hills pack roams an area from southern Bayfield County near the Sawyer County line, to north across Highway M, portions of the Namakagon River, Pioneer Road, and to the southern edge of the Porcupine Lake Wilderness Area. The pack roams from western portions of Lake Namakagon to just a few miles east of Cable, Wisconsin. The Seeley Hills pack has the northern Birkie trail run right through its territory."[14]

Adrian went on to hypothesize that "moves to the east side of the territory by the pack, including the Rock Lake area, may have been partly due to all the skiers in town for the American Birkebeiner ski race, and lots of people in the western parts of its territory last weekend."[15] That makes sense to me. That's exactly why I was skiing at Rock Lake too! With a love of wild places, a tolerance of cold, a taste for venison, and an aversion to crowds, I have quite a bit in common with the wolves who share my woods.

The Black-Masked Bandit

Brad Gingras

A colorful line of 30 students, several parents, and a couple of teachers—all on snowshoes—stretched out behind me on the Ridge Trail at the North End Trailhead near Cable, Wisconsin.

The students could sense that the hike was almost over, and I'd already reassured some tired souls that the trail was all downhill from here. I was looking ahead to my next interpretive stop and eavesdropping on the students' excited conversations behind me when a teacher called out, "Hey, Miss Emily, there's a dead mouse in this tree!"

Well, that stopped me in my tracks. "Are you serious?" I called back, nearly bursting with excitement. "That's awesome!" I stepped out around the students and started tromping through the deep snow back to where she stood. I waved to the kids at the front of the line to turn around and walk back with me.

There, nestled in the delicate fork of two sugar maple twigs, just above my eye level, was a very dead mouse. "That's awesome!" I said again, probably striking fear into the hearts of the parents who were wondering how someone this weird came to be leading their children through the woods.

And then I explained: This mouse was most likely stuck here by a bird called a northern shrike. When shrikes are able to catch more food than they need in a day, they store it for later use. Sometimes they'll impale prey on thorns or barbed-wire fences. When those aren't available, the fork in a twig will do.

Stored prey provides the shrike with food security and will eventually get eaten when the hunting is poor. A male shrike with abundant prey impaled throughout his territory has a better chance of attracting mates and fathering successful nests, but this bird probably wasn't worried about attracting mates. Breeding takes place north of 50 degrees latitude around the globe (that's northern Canada). In winter, shrikes migrate only as far as necessary to find food, which means they often come to Wisconsin.

Surprisingly, these skilled predators are songbirds. Being songbirds, shrikes lack the sharp talons of raptors like hawks and owls. Being songbirds, shrikes have another weapon. Like the winged Sirens of Greek mythology, shrikes sing sweetly to attract other songbirds.

Once prey are lured in, shrikes attack with a solid blow,

then finish the job by biting the neck, shaking, or repeated knocks to the skull with their sharp beaks. Impaling prey on thorns or sticking them in forked twigs may seem brutal, but it is also a practical way to compensate for having delicate songbird feet that cannot grip food during dinner. Sporting a black bandit mask on their gray heads, northern shrikes look the part of a feathered villain.

More than half of a shrike's diet is small rodents like mice and voles. Right now, most of those tasty little critters are safely hiding beneath a foot of snow in the subnivean layer. This unlucky mouse must have been caught running on top of the snow, above the relative safety of the subnivean zone. While foxes and owls have enough mass to break through the crust and dig for tunneling mice, shrikes do not have that ability.

We hadn't seen a single mouse track on the snow in our two days of field trips, so this mouse might have been caught and stored a few days ago, before the last storm. It looked a tad dehydrated.

After snapping a couple of photos, I snowshoed back to the head of the line and kept the group moving. I'm not sure what those kids (or teachers and parents) thought of our discovery. A gross oddity? A cool animal sign? A scientific curiosity? In any case, the naturalist in me is thrilled to find evidence that a shrike is visiting our woods. I'm not the least bit sad that one less mouse is running around, especially since it came to such an educational end.

Dalaney Rasmussen, age 10

SPRING

A Prickly Botanist

Ayla Baussan, age 8

A dozen adults, hunched over tables, were examining bare twigs with borrowed hand lenses. Back and forth they looked between the dichotomous key and their specimens. Did it have one or three bud scales? Or more than five bud scales? Was the pith of the twig chambered or solid or hollow? Finally we arrived at an answer: the smooth, gray twigs with plump, red buds (covered by three bud scales) belonged to a basswood tree. For fun, I had the students each nip off a bud with their teeth and experience the slightly sweet flavor and mucilaginous texture of this wild edible.

We practiced keying out a few more twigs, and then the winter tree identification class eagerly headed out into the warm sunshine. We sniffed the glossy, aromatic buds of balsam poplar; admired the snazzy, scarlet stems of red osier dogwood; and commented on the stalked buds of speckled alder. Identifying woody plants by their winter twigs is one of my favorite challenges, and I hoped the students would gain an appreciation for it too.

As I drove home a dark blob high in a roadside tree caught my eye. One quick glance backward confirmed my suspicion. The blob showed a spikey outline against the evening sky. Porcupines have begun their spring feast!

If you want to meet a real expert in winter tree identification, be sure to bring a thick leather glove for shaking their paw. Porcupines survive on low-nutrient bark all winter long, and the lengthening days of spring bring a welcome change in their diet. A few years ago I started noticing that in late winter these bristly roadside silhouettes begin to appear with some regularity. As it turns out, porcupines spend their spring eating a rotating smorgasbord of juicy tidbits, and they time their courses to match the period when each entrée is the most nutritious. They know their trees well, because their life depends on it.

Dr. Uldis Roze, professor emeritus at City University of New York and author of *The North American Porcupine*,[1] radio-collared a porcupine he named Rachel and followed her around the Catskills one spring. By testing the chemical content of the buds and leaves while Rachel fed on them and then analyzing the plants again after she ceased to eat them, Roze discovered that Rachel pinpointed plant parts high in protein and low in toxins.

To get at the most nutritious parts of a twig, porcupines will balance out toward the terminus of a branch and, using their self-sharpening incisors, nip off its end. Turning the wand around, they nibble off all the most tender twig tips and buds and then discard the rest. Trees on sunny roadsides may be farther along in their seasonal sequence than their relatives in a shady forest, so porkies are extra visible now as they satisfy their hunger for spring.

Sugar maple buds are one of porcupines' first spring snacks. As sap rises, the buds swell. Packed inside the overlapping layers of toffee-colored bud scales, tiny leaves and twigs begin to expand. Their chemistry changes too. At the height of their nutrition (which happens just as the syruping season ceases), sugar maple buds have more protein than enriched breakfast cereal.

After a few weeks, once the buds have fully released their tender young leaves, porcupines cease their sugar snacking totally. The protein content might still be the same, but it no longer matters. The tree has now fortified its appendages with toxic tannins. Tannins are chemicals that bind with proteins, making them an effective defense against many types of herbivores.

Sugar maple's close cousin, the red maple, defends its

buds, twigs, and leaves with tannins year-round. It doesn't seem to have a place in the porcupine's buffet. Instead, the sweet basswood buds that the students tasted are a porcupine delicacy, as are expanding aspen and willow catkins.

Ash trees are the last to open their buds, and they also represent that last of the spring smorgasbord. Once the juicy expanding packets of ash leaves get too old, porcupines switch to a diet of ground plants for a couple of months. When the dandelions, clover, and raspberries dry out, porkies head back up into the trees and return to their winter diet of tree bark and evergreen needles.

Phenology is the study of when specific events happen in nature from year to year in a specific place. Each tree species has its own place in nature's calendar. Porcupines follow this schedule closely, since each species has an optimum nutritional stage. Amazingly, porcupines are active mostly at night, and so they identify their next meal by bud and twig under the cover of darkness. Perhaps my next tree identification workshop should include a pricklier instructor!

Hot and Cold for Spring

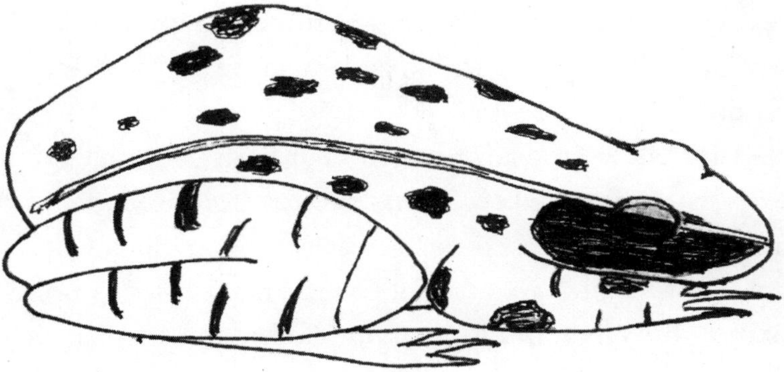

Helen Bolgrien

During the fickle days of early spring, my emotions seem to rise and fall in sync with sap in the maple trees. I get kind of grumpy when the mornings dawn damp and cool and the tracks on the ski trails sink dejectedly into the slush.

My feelings change as the days brighten, intense sun takes over, and the temperature spikes. Warm breezes and dry roads make me eager to put away the skis and happy to pull my road bike out of storage. I guess you could say that I'm running hot and cold for spring.

The first place I head on my inaugural spring bike ride is to a little wetland across the county road from Lake Namakagon. This wetland tends to thaw a bit sooner than the one by my house, and it is consistently the first place I hear the spring chorus of frogs.

The eleven species of Wisconsin frogs (and one species

of toad) begin singing their songs in a typical order each spring. Their specific phenological timelines depend on things like cold tolerance, overwintering location, and breeding habitat.

Our three earliest singers share some common characteristics. Wood frogs, chorus frogs, and spring peepers are all somewhat terrestrial frogs that spend the winter as "frogcicles," buried lightly in duff on the forest floor. Spring sun and warm rains can reach them quickly here, triggering them to thaw. Since they wake up next door to their honeymoon suites, they can breed early in the season. April is typically their main month of romance, but each spring progresses a little differently.

By getting such an early start, these frogs are able to carry out their entire breeding cycle in vernal pools. These woodland puddles persist for only a few months each spring. This means that eggs and tadpoles don't have to worry about the fishy predators who live in permanent bodies of water, but it also means that the parents need to hurry up and get their babies growing so that they can sprout legs and hop away before the pool dries up.

Of course, spending winter on land has its own challenges. Even though the subnivean layer—where snow meets earth—remains warmer than the television weather report indicates, frogs in the duff must still endure below-freezing temperatures for many days at a time. They do this not by avoiding the frogcicle state but by guiding it.

Wood frogs are the most highly studied, but spring peepers and chorus frogs use similar techniques. The process starts early in fall, when wood frogs begin accumulating urea in their tissues. While urea is a waste product that humans excrete in urine, urea is also a cryoprotectant, or a substance used to protect biological tissue from freezing damage.

When temperatures dip below 32 degrees, ice crystals form on the frog's skin. Some animals are able to "supercool," which means that they prevent the formation of ice even when temps dip below freezing. This requires them to avoid contact with ice that would trigger flash freezing. Frogs, with their permeable skin, allow the frost to enter and set events in motion.

Ice formation causes the frog's liver to convert stored glycogen into glucose. The sugar floods the frog's body, carried in its bloodstream by a heart that keeps beating as long as possible. As ice forms outside cells, it locks up the pure water and leaves behind a higher concentration of salts and other things that had previously dissolved in the liquid. Water from inside the cells wants to flow outward, across the cell membrane, to even out the imbalanced concentrations.

This could dangerously dehydrate the cells, though, and there is a threat of them collapsing. Glucose and urea help the cells retain water in two ways. First, sugar water freezes at a lower temperature than plain water, and so less ice forms overall. Second, the increased concentration of chemicals inside the cell keeps water from flowing out of the cell under osmotic pressure.

While frozen, the frogs are essentially dead.

Once spring comes, however, wood frogs thaw from the inside out. The heart starts beating (scientists still don't know what jump-starts it), the brain wakes up, and finally the legs move. Those legs carry them directly to a nearby vernal pool. A deafening chorus of love songs ensues, and mating commences in a hurry. Wood frogs quack, spring peepers jingle like bells, and chorus frogs imitate a finger plinking a plastic comb. That's what I was hoping for today, but the wetland was still quietly blanketed in snow.

Over the winter the frogs may undergo multiple freeze-thaw cycles. Springtime warmth also ebbs and flows. And then there's me, sometimes turning a chilly shoulder to the next season and sometimes embracing it with an open sweater. I guess I'm not the only one who runs hot and cold in spring!

The Pulse of a Waking Forest

Kate Stolp

The morning dawned crisp and blue. In the woods I knew that the trails would be firmly frozen, the mosquitoes still far from flying, and perhaps the ticks would be hunkered down too. The novelty of this snowless, chilly spring weather was a perfect match for a hike on the North Country Trail.

As I entered the brushy, deciduous forest and strode down the leaf-littered path, I felt my heart quicken and thud in my chest. After a moment, the sensation reached my ears too. It wasn't my heart—it was the drumming of a ruffed grouse! I'm always amazed by how much I feel their sound instead of hearing it. In the thick of mating season, their incessant, accelerating beats have even caused a fleeting catch in my breath as they temporarily overpower my own body rhythms.

The grouse drummed again; thumping slowly at first and then crescendoing into a rapid-fire blur. As his last vibrations dissipated into the still air, another grouse answered from a neighboring territory. The low frequency sounds are audible from up to a quarter mile away.

Several explanations of how ruffed grouse create this sound have been employed over the years. Because male grouse often display from atop a hollow log, perhaps they create the sound by hitting the log with their wings. Or, since they also display on rocks, mounds of soil, or prominent roots, perhaps the sound comes from the wings striking together behind the bird's back, like a spruce grouse. Those explanations were disproved in part due to some fuzzy photographs of captive ruffed grouse taken by Professor C. F. Hodge of Clark University in the early 1900s.

H. E. Tuttle, a wildlife artist and naturalist, spent many days in blinds observing drumming grouse from 1910 to 1918. Tuttle also examined Hodge's photos and determined that the sound did not come from wings beating together. He posed the possibility that rudimentary air sacs contribute to the sound (as in the displays of greater sage-grouse).

One theory he dismissed heartily was that the drumming sounds were produced the same way as grouses' noisy flight. He described it as "an unsatisfactory explanation of that far-away throbbing challenge which steals on the ear so subtly, like the half heard beating of one's own heart."[2]

As in many realms of science, as the technology improved, so did scientific explanations and accuracy. In the 1920s Arthur Allen of Cornell University used a new-fangled contraption to shoot motion-picture footage of the grouse. By slowing down the movement, he ruled out every explanation except the one accepted today.

On the Cornell Lab of Ornithology's "All About Birds" website, you can read the result of their founder's research: "The male ruffed grouse's signature drumming display doesn't involve drumming on anything but air. As the bird quickly rotates its wings forward and backward, the air that rushes into the temporary vacuum beneath the wings creates a miniature vacuum, generating a deep, thumping sound wave."[3]

That thumping sound wave sends "the blood sap pulsing quicker along the veins" (Tuttle again). This works not only for humans but also for the lady grouse. She can probably differentiate between males, since the number and rate of pulses in each bout of drumming is unique to each individual.

Tuttle noted that if he rustled the leaves in his blind to sound like the dainty footsteps of an approaching female grouse, the male would drum instantly, and also flare his namesake ruff of neck feathers. Once he's able to attract a female with this huge output of energy, copulation lasts only a few seconds. The female then wanders off to build a ground nest and raise the chicks completely on her own.

Hiking down the trail, I found many clusters of sawdust-filled grouse droppings. Those piles marked where they had digested tree buds and catkins on subzero nights while buried snugly in a snowdrift. In one balsam thicket I heard—but did not see—the whirr of a grouse's startled explosion into flight. I have to admire the tenacity of these year-round residents, even if I sometimes chuckle at their lack of grace.

When the days lengthen with the promise of spring, I appreciate how grouses' drumming seems to jump-start the pulse of a waking forest.

Northwoods Monkeys

Norah Werner, age 7

Sunlight streamed through the tree trunks as I meandered into the warming woods. The ruffed grouse's thumping faded into the distance as I left his territory. In its place a loud, staccato drumming echoed through the hills, followed by the wildly laughing cry of a pileated woodpecker. My ornithology professor nicknamed pileateds the "monkeys of the Northwoods" due to their raucous, whinnying calls.

Like grouse, pileated woodpeckers are not daunted by northern winters and maintain their mates and territories throughout the year. Activity has ramped up recently with the onset of spring and breeding season. Unlike the melodious warblers who are singing their way north right now, woodpeckers use drumming to attract a mate and defend a territory. "With such energy did he hammer that his whole body shook and his wings quivered. He fairly hurled himself wildly at it," described the naturalist Ernest Waters Vickers in 1915.[4] They add to the pulse of the forest.

Woodpeckers beat on trees for three main reasons: to forage for food, to excavate a nest cavity, and to drum for communication. The first two are relatively quiet affairs, done slowly and deliberately. The third is as loud as possible. Woodpeckers prefer the resonant acoustics of a hollow trunk or branch on which to drum—or a stovepipe or rain gutter—to make sure their message gets across loud and clear. Both male and females drum.

I'm not sure who was drumming the beat I heard, but as I peered through the tree trunks, one large bird swooped through the maze and landed low on a bole, followed shortly by its mate. Then again one swooped off and the other followed. Were they searching for food together? Or performing their mating dance? In 1908 Francis H. Allen witnessed their ritual:

> They hopped up and down the trunk, frequently pecking at each other's bills simultaneously, now on one side of the tree, now on the other . . . They hopped backward and downward a great deal, and often they lifted and partly spread their wings. Their

> motions were limber and undulating, marked by a certain awkward grace.[5]

After the thrill of their mating rituals come the more domestic duties. Pileated woodpeckers use their large, straight, chisel-like bills to excavate a new oblong nest hole each year. The male seems to do most of the excavation. Dense, mature forests tend to contain the large, dead trees that they prefer. In young forests, when old, dead nest trees are the tallest ones around, lightning can present a hazard to nesting birds.

Although a significant time investment is required—each cavity takes from three to six weeks to complete—woodpeckers will not reuse the hole in subsequent years. This probably serves to protect the babies from parasites or from mammalian predators who might remember the location from year to year.

In the years after the woodpeckers raise their brood of three to five young, a wide variety of birds and mammals will use the abandoned cavities. Other woodpeckers, wood ducks, bluebirds, flycatchers, owls, bats, squirrels, and pine martens all benefit from the predrilled cavities. This crucial role in creating habitat makes pileated woodpeckers a keystone species.

The pileated pair leapfrogged their way deeper into the forest, and I continued hiking. Several times I noticed piles of fresh wood chips near the trail and looked up to find a tree trunk perforated by large, rectangular holes from the woodpeckers' foraging activities. Nest trees have no such telltale signs, since the birds carry away all chips, nest debris, and feces that might give away the location of eggs and babies to a nest predator.

These days it seems pileated woodpeckers are having good luck with reproduction. Hardly a day goes by when I don't hear the "monkey of the Northwoods," or catch a glimpse of this striking black and white bird in undulating flight. It wasn't always that way. Pileateds were considered rare in the 1900s due to habitat loss and hunting. When the forests rebounded, so did the birds. I like the sound of that.

Turkey Vultures

Linda Luhman

About a month ago, in mid-March, my dad reported seeing his first turkey vulture of the year down in Iowa. Most people probably don't associate these drab, brownish-black scavengers with spring—or even realize that vultures may fly as far as South America for the winter—but they are one of the earliest returning migrants.

Sure enough, a few weeks after Dad's report, I caught sight of the V-shaped wings and rocking, unsteady flight of a turkey vulture soaring above my road. I suppose the "buzzard" had to wait for carcasses to thaw and start sending wafts of putrid odors skyward.

For as long as I can remember, Dad has been pointing

out every turkey vulture (TVs, he called them) circling over every road trip we ever took. Those gray-fingered wings are distinctive and built to soar. It was fun, even as a kid, to easily identify such a large bird flying so high up in the sky. Plus they have an excellent gross factor that kids love. "Don't get too close to a turkey vulture," warned the park ranger on my first grade field trip to Effigy Mounds National Monument. "They'll throw up all over you!" That I still clearly remember this fact, and the moment I learned it, is an excellent testament to the power of outdoor education.

Since that day I've discovered many more disgusting facts about turkey vultures. First of all, projectile vomiting is a defense they use against predators, not just curious humans. The foul-smelling mix of semidigested meat and digestive fluids can sting if it reaches the predator's eyes. In addition, emptying their stomach may be necessary to lighten the load for takeoff and escape if a TV is interrupted while gorging on a roadside carcass.

Turkey vultures don't just spray nasty stuff on enemies; they also defecate on their own legs. This habit has a scientific name (urohidrosis) and a valid purpose. As water evaporates from the combination of urine and feces (birds don't separate their waste like humans do) it cools the blood vessels in their legs and feet. It's quite handy, actually, since the birds can't sweat. Plus, the urine helps to kill bacteria they might have acquired while walking over a dead animal.

Even without sweating, vultures' feathers sometimes

become damp during dewy, foggy, or rainy nights. Then, while they wait for the air to warm enough to begin rising in thermals, TVs perch in a spread-winged stance in the sunshine. This not only dries feathers but also warms their bodies and bakes off bacteria.

Not everyone sees vultures as gross. Tibetan Buddhists practice sky burials, where vultures consume the dead, and Buddhists respect the birds for their role in cleansing the earth and continuing the food chain.

While their diet of rotting meat may repulse us, turkey vultures dispose of carcasses that could otherwise breed disease. TVs have developed excellent immune systems that can ward off and even destroy the microbes that cause botulism, anthrax, cholera, and salmonella. Can you imagine a world in which dead things all rotted slowly in place? Their stomachs, icky as they may seem, help purify the world.

Indeed, despite their gross appearance, every adaptation of the turkey vulture seems aimed at cleanliness and purification. Even their bare, ugly heads serve to keep turkey vultures clean. While they are feeding on carcasses, TVs sometimes need to stick their heads deep into the body cavities of the dead animals. They preen the rest of their body feathers frequently, but can't clean their own head. The lack of feathers allows sunlight to sterilize their skin.

So although they may appear gross, TVs' scientific name—*Cathartes aura*—is quite appropriate. It means "purifying breeze."

And in fact, the story of the vultures—of winter's rotting wounds transformed and purified, of the purifier rising up into the sky, of it returning to cleanse the world again and again—sounds a lot like another story I often hear about in the springtime.

The Symphony of Spring

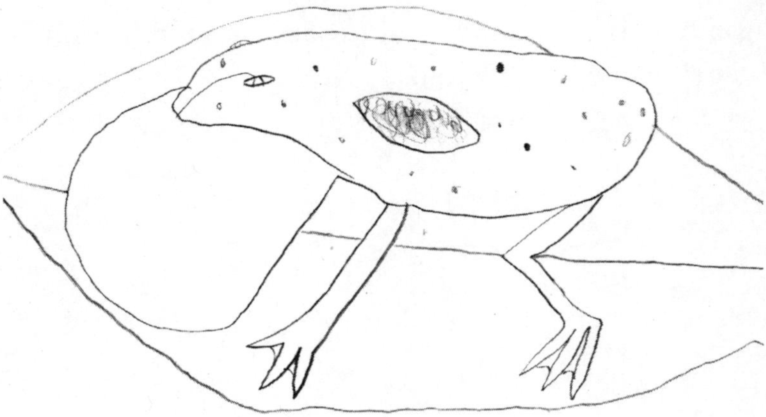

Wyatt Strapon, age 8

A big truck rumbled by as I squeezed the brake levers and guided my bike onto the shoulder. The wonderfully warm, sunny afternoon had lured me out on an adventure, and my first stop was a small, roadside wetland. With a southwest exposure, this marsh thaws a day or two sooner than the one by my house. Surely I would hear some frogs.

As the truck's noise faded, silence fell over the forest. Peaceful, yes, but not what I'd hoped for. Could the wood frogs still be waking up? It does take them a little while to recover from being frozen all winter. With the arrival of warm temperatures, wood frogs thaw from the inside out.

Recovery is relatively rapid—especially considering what

they've been through—but not instant. The frog's heart starts beating before they are fully thawed. Breathing resumes soon after, and circulation begins as soon as their blood melts. That doesn't mean the frog is ready to hop off into the sunset, however. It takes more than five hours for their leg muscles to regain some function and up to twenty-four hours for the frogs to exhibit normal body postures and coordinated muscle functions.

A couple of days later their mating drive kicks in. That's what I stood there hoping for in the silent woods. Then from the back corner of the wetland came a soft quack. And another. Then a second frog called softly from another shore, and a third chimed in tentatively from near the road. From there, the chorus crescendoed until a cacophony of quacks, chuckles, and clucks echoed through the woods.

For the next few weeks, this wood frog choir will continue its performance whenever the temperature rises near 50 degrees and above. Mostly males call, and the impassioned sound serves to attract females to their pond. The more females their calling can attract, the better chance each male has of finding a mate.

As frogs arrive at the pond and mill around, eager males will search for a receptive mate. They can't see or smell a difference between females and other males, though. Instead, a male frog hugs another frog from behind and around its waist. Fellow males, identified by their slim waist, give a loud croak and are released. A slender female, who has already released her eggs, also gets a pass. But when a male frog grips a plump female—chock full of eggs—he hooks his thumbs around in front of her and doesn't let go. This is called amplexus.

From there, the female deposits her eggs in a floating mass, and he fertilizes them as they emerge from her body. Throngs of wood frogs in a pool may lay their eggs together. The earliest breeders benefit from their eggs being in the center of the mass where they absorb heat, develop faster, and gain protection from predators. Each frog's four-inch diameter egg mass contains from 1,000 to 3,000 eggs.

Eggs laid in the cold infancy of spring will mature slowly, over the course of a month. Later in the season when the water has warmed, the eggs may hatch in only 10 to 14 days. The amphibious parents have completed their responsibilities, though. They don't stick around to provide care, cheer at baseball games, pay for dance lessons. . . or teach their little tadpoles to ride a bike. I'm thankful I'm human!

As the frog chorus reverberated, I swung back onto my bike and continued around the lake. Wood frogs were only the beginning.

With my ears now tuned in, the jingle bell chorus of spring peepers became apparent everywhere. A bright orange fox with sleek new fur bounded through the ditch. The clear whistle of a song sparrow reached my ears. All along the road, frogs croaked and peeped, robins whinnied, juncos trilled, and geese honked. Silence has gone out of style. Let the symphony of spring commence!

The Loons Are Back

Suzanne Stone

Cold water and rough sand swirled around my toes as I stepped into the lake. My cabin-fevered feet welcomed its caress. With one last heave, I slid my goldfinch-yellow kayak into the water. Frothy gray clouds left over from scattered showers polished each ripple into a mirror as I pushed off from shore.

Easy strokes soon propelled me around the point, heading for the bay where the loons always nest. Are they back? I've been expecting to hear their wails slip in through the tiny crack I've opened in my bedroom window. But so far only a chorus of frogs has been shouting my lullaby, and the calls of other returning migrants have filled the days: the tiny, urgent winter wren letting forth a tinkling stream of notes in the ravine; tireless song sparrows shouting "Maids, maids, put on your tea kettle-ettle-ettle" over and over from the highest tips of alder brush; one white-throated sparrow singing

"Oh sweet Canada, Canada, Canada" as it hurried north. I have been hearing the prehistoric rattling cries of sandhill cranes and the whimsical whinny of robins—but no loons.

Scanning the waves for any speck of a life, movement in the tree line caught my eye. I strained for a better look. Great flapping wings, with both white and black feathers, came in and out of view. "Could it be?" I hoped. And then I nodded with satisfaction as the great shape swooped up to land on the utility pole where the ospreys live. Last year that nest ended in tragedy, with only one out of the family of four surviving. This year, new owners have already moved in and their nest is half-built.

Turning into a boggy bay, I startled some wood ducks, who burst out of the dried reeds in a flurry of squeaky alarm calls. Then another noise—or was it a movement—caught my attention. Either way, I turned back toward the osprey nest in time to see a flash of white on an eagle's head as it wheeled in the sky, chasing—or being chased by—the smaller osprey.

Eagles are majestic. They are also bullies. The technical term is *kleptoparasites*, which means that they will bother an osprey until it drops a fish, then swoop down and claim the snack for their own. Eagles are also the only common predator of adult loons—which is what triggered the sound that caught my attention. Two loons were alarming at the eagle with harsh cries to send shivers up the spine. "They are back!" I thought as I alternated between scanning the water and following the aerial battle. A smaller bird, probably a crow, joined the chase. Soon the eagle was on the run, and the osprey escorted it to some invisible boundary before swinging around and heading back to its nest.

I paddled toward the now quiet loons. Straining my eyes, two black dots soon distinguished themselves from the distant shoreline. It had to be them.

At this point in the season, any two loons who can be in such close proximity without yodeling conflicts must be a pair. Are these the same loons who reared one chick safely into October last year? It is impossible to tell without the loons being banded, but the research suggests that it is likely. Loons have an incredible sense of place. They will choose a territory as much like the one they were born on as possible. Once they have a territory, they prefer to keep it. The place-based knowledge they acquire each year—the best nesting spots, best fishing spots, common hazards—is invaluable. While mating for life is possible for loons, their attraction to a territory—not each other—is what keeps them together.

Soon I had slid close enough to see the gleaming white and inky black checkers on their backs and the sleek swoops of their foreheads. Over the winter, loons lose all of their flight feathers at once as they molt back into breeding plumage. It is known as the "pre-nuptial" molt, and they arrive here looking spiffy and ready to woo a mate.

I drifted in the evening light as the loons courted nearby. The birds swam toward each other for a distance. When they got within several feet, one loon spun around and they both swam the same direction. After a moment, one dove, surfaced farther away, and the routine would start again. Their coy behavior and snazzy black and white garb reminded me of a note I read about the senior prom in the local newspaper's events calendar.

Spring is here. With it comes young and hopeful life decked out in many styles of tuxedoes and rainbows of colorful gowns. Everyone must sing, perform, defend, watch, or choose according to their own individual natures. Who have you been admiring this spring?

Queens of Spring

Ella Mae Richardson, age 11

The gentle sound of waves lapping at the shore tugged me toward the lake. The last sheets of ice in my little bay had finally melted away under the persistent caress of a warm breeze. Late afternoon sunshine prickled the back of my neck. A dead branch lay across the trail, nestled in the moldering leaf litter, victim of the winter winds. As I stooped to pick it up, a bit of color caught my eye. Zooming low above the moss was a big, fuzzy bumble bee.

I didn't react quickly enough to stop my momentum, so the flying branch scared off the flying bee. I did get a brief glimpse of a bright orange stripe in the middle of the classic black and yellow body.

I've seen a bee like this before as it wallowed in the bounty of summer on a dazzling purple flower. I eventually identified it as *Bombus ternarius*, an orange-belted bumble bee. The landscape on this April day, although warm and sprouting, was a far cry from that flower-filled August field. Most bees don't fly this late in the day, even in summer.

Bumble bees are able to fly at colder temperatures than many other insects. Flight muscles don't work when they're too cold, but the temperature limits of bumble bee flight are lower than most. In addition, they raise their body temperature above the ambient temperature using a few neat tricks.

First, bumble bees readily absorb what little sunshine peeks through the spring clouds. Then they generate their own heat by shivering their muscles without flapping their wings. It can take five minutes of shivering to raise their muscle temperature to the required 86 degrees when the air temperature is 55. Finally, the thick fuzz that covers their bodies helps retain their hard-earned heat.

All these adaptations mean that bumble bees can be active earlier in the spring, earlier and later in the day, and in more adverse weather conditions than honey bees and other not-quite-so-hairy bees.

Still, this early in the season, I must have been in the presence of a queen. Each fall the old queen bumble bees, the workers, and the male bees die in the cold, while newly fertilized queens find a safe underground place to overwinter. When they emerge in the spring, the queens are on their own for at least five weeks until the eggs they lay become a hard-working colony.

While one may imagine that all bees nest in hives or hollow trees like the European honey bees that farmers use to pollinate crops, native bumble bees tend to build their colonies in abandoned mouse nests or other underground burrows. The burrows are sheltered from the weather and preinsulated with grass, thistledown, and mouse fur. Many other types of native bees are solitary and don't form colonies at all.

It takes a lot of fuel to keep a big bumble bee buzzing, so the queens will often try to locate their nest near a food source. On this lakefront surrounded by hemlock forest, the only flowers I'd seen were on willows (an important source for early spring pollen), but elsewhere, patches of Dutchman's breeches are said to attract the bees. Those oddly shaped white flowers are consistently some of the first blooms I spot in the woods.

Not just any insect has the strength and tongue length to forage for nectar in these little pantaloon-shaped flowers, though. The queen bee grabs on to one of the white flowers, and the one next to it too, for stability. Then she'll first burrow her tongue and then her head into the flower and angle toward one of the two nectar spurs. Throughout this process, the flower's anthers brush pollen onto her head, thorax, and forelegs. Smaller bees can't quite pry open the petals to access the juice bar inside, and they sometimes cheat by chewing holes in the nectar spurs.

While the plant's desire is for the pollen to travel to the female parts of another flower of the same species, the queen bee uses some of it—and nectar too—to fill a tiny honey pot inside her burrow. This food will sustain her while she broods her nest. When the pantry is stocked, she lays her eggs. After four days the eggs hatch and she feeds the larvae on nectar and pollen while they grow. After two weeks as larvae and two weeks as pupae, the queen's daughters emerge as adults—full grown and ready to work. Her hard work has paid off.

Queen bumble bees are absolutely essential pollinators in early spring, but throughout the year Wisconsin's 13 species of native bumble bees contribute mightily by pollinating native plants and also crops like tomatoes, peppers, squash, cranberries, and blueberries. Their capacity to fly in cooler temperatures and ability to muscle into certain flowers make them invaluable.

Bumble bees are not invulnerable, though. Threats from parasites, pathogens, pesticides, invasive species, climate change, and habitat loss, among others, challenge their survival. There's quite a buzz around bees these days, and it's not hard to understand why. They are truly the queens of spring.

Kingfishers

Brendan Bower, age 7

Fresh sunshine warmed my bare arms and legs as I biked along the eastern shore of Lake Namakagon. The joy of speed, freedom, and my trusty steed brought on an irrepressible smile. As I neared the bridge over a marshy flowage, a cacophony of birdsong began to cut through the whirr of tires on pavement and wind in my helmet. I pulled onto the shoulder for a better look.

Red-winged blackbirds gurgled and trilled their early spring song that is so energetic it seems to cause ice to melt and streams to flow. A pair of Canada geese honked harshly from their grassy hummock, building their pair bond by shouting together. And then a rattling cry ripped through the air. My head snapped up automatically as I followed the sound and the blur with both my ears and my eyes. My focus finally landed on a slate-blue shape just settling on the branch of a broken snag.

"Kingfisher," I whispered, as I whipped binoculars out of my bag. Recently back from Central America, his breast was pure white with a jaunty blue band to match his powder-blue tux. A female would have had a beautiful rusty bar across her belly. Belted kingfishers are one of the few bird species where the females are more colorful than the males.

In the next moment I realized he had something in his mouth—a large minnow, probably—and he was whacking it against the branch over and over with sideways jerks of his head. At some point, when he decided it was sufficiently dead, he tipped his head back, gave the fish a little toss, and down the hatch it disappeared!

While this minnow was comfortably bite-size, kingfishers, who are thirteen inches long at most, have been known to eat fish up to seven inches long. Their trick? They leave part of the fish hanging out of their beak until stomach juices digest enough of the fish that it can slide all the way in. Also, the bird's tiny tongue stays out of the way.

Young kingfishers have such strong stomach acids that they can digest fish and crayfish completely—bones, shells, and all. Once they leave the nest, kingfishers regurgitate pellets filled with the indigestible bits, just like owls.

Those young belted kingfishers, who hatch in May or June, keep their parents busy by consuming their body weight in food each day. Luckily, kingfishers live up to their name with the help of several adaptations.

First, they can scope for prey either from a poolside perch or by hovering in one place high above the water. Like many birds, kingfishers have excellent eyesight. Oil droplets on their cone cells enhance their color vision and reduce glare as they search for prey. Two focal points in their eyes allow them to adjust for the change in refraction angle between air and water as they dive after a fish. The position of these two foveae gives kingfishers monocular vision in the air (each eye is used separately) and binocular vision underwater (which allows them to judge distance). A third eyelid, called a nictitating membrane, acts like protective goggles.

When a kingfisher dives, its sturdy, conical beak wedges into the water and pushes it aside with minimal impact or splash. Japanese bullet trains, which operate at speeds exceeding 200 miles per hour, imitate the aerodynamics of the kingfisher's beak to reduce noise pollution and improve fuel efficiency.

After grabbing the fish or crayfish with its viselike bill, the kingfisher swoops up to a perch to whack away, like I just witnessed. The young are fed small fish or regurgitated food. As they grow up, parents teach kids to hunt by dropping fish into the water for the students to retrieve, just like in common loon families.

While their hunting abilities are impressive, a

kingfisher's strangest adaptation is used for digging nest burrows in sandy embankments. Their inner two toes are fused together, resulting in what appears be a single, flattened toe. This trait, called syndactyly, is shared by kangaroos, wallabies, opossums, koalas, and wombats.

I will never cease to be amazed by the ways that nature has sharpened its every skill and perfected its many tools for survival. As I hopped on my bike (a wonderful product of human adaptation and innovation) to continue the ride, the kingfisher sent me off with one more rattling cry.

An Ephemeral Mystery

Donna Post

On one of those damp, gray days in early spring, I joined the North Country Trail Navigators to hike a brand new section of trail east of Copper Falls State Park, near Mellen, Wisconsin.

Light was low, but our spirits were high. Many of these women completed the 100-mile challenge on the North Country Trail for the National Park Service's birthday in 2016. Now they are addicted and trying to hike this year's miles before the mosquitoes hatch.

From the gravel road where we parked, the trail snaked its way through a beautiful sugar maple forest. The rich soil, typical of maple forests, was soft and loamy: great for plants, but not for hiking boots. We giggled, whooped, and exclaimed as we slipped and slid through mushy spots. With use, the soil will compact and become a more durable tread, but ours were some of the first boots to travel here.

I was at the end of the line chatting away when I heard my name called up ahead. One of the women had spotted a mystery near the trail. In several patches, each about the size of a manhole cover, the thick maple-leaf duff was pushed away. Laying exposed on the black soil were scattered chains of yellow-green plant rhizomes (horizontal stems) that reminded me of the little plastic pop-together necklaces I won at carnivals as a kid. Each segment of the rhizome was an oval-shaped "bead." Also visible was the lacy network of regular, thin roots.

Of course, with a naturalist along, the group wanted an explanation of what we'd found. I pulled out my camera to photograph the creeping rootstocks, but I couldn't put a name on the discovery. A few tiny clusters of curled-up leaves sprouted from the ends of the segments, but they weren't big enough for a confident identification. I listed off a few possibilities: blue cohosh, trout lily, spring beauty, trillium, Dutchman's breeches, wood anemone . . .

One thing I was certain of was that these roots belonged to a spring ephemeral. This category of flowers completes their entire life cycle in the two months or so between snowmelt and leaf-out. Instead of adapting to grow in the oppressive shade of the summer forest canopy, these little plants take advantage of the full sunshine of early spring.

During their brief growing season, spring ephemerals bury an energy reserve in tuberous roots, modified underground stems, or bulbs. Those storage units lay dormant—and protected from drought or cold—through the winter. Come spring, leaves and flowers erupt quickly with the plentiful moisture and sunshine. They provide food for early pollinating insects, and then restock their provisions before melting away under the deepening summer shade.

Because so many of the spring flowers that inhabit maple forests have similar strategies, the roots we found could potentially belong to any number of the beloved blooms that grace each spring. To me this was the most interesting part of the mystery. I did, of course, do some research at home to try and match these particular roots with a name. Finding good, identifiable photos of roots is hard, though. Happily, the challenge just meant that to solve the mystery with confidence, I would have to return to this beautiful trail once the leaves of these cryptic plants had expanded.

My chance came two weeks later. The trail was only slightly drier, but the scenery had gone through a delightful transformation. Little green leaves pushed up everywhere, but mostly they sprouted out of the trail itself, where dark soil had warmed more quickly than the pale duff. Trout lilies, trilliums, leeks, oh my! And then I noticed a small patch of those mysterious, yellow-green segmented roots. Erupting out of their tips were the miniature palm-tree-like leaves and clustered flower buds of cut-leaved toothwort. Mystery solved.

Cut-leaved toothwort (*Cardamine concatenate* [new] or *Dentaria laciniata* [old]) is a widespread spring ephemeral in the mustard family. Its common name describes both the narrow, toothy leaves and the canine tooth-shaped sections of it rhizomes. The pretty white flowers in its elongate clusters will each have four petals. True to its family, the leaves and the rhizomes of toothwort have a spicy taste, similar to wasabi. The flavor probably deters some herbivores, but it's encouraged people to use it as a condiment and a medicine over the years. I chewed on a leaf, and it tasted like a very plain mustard.

One herbivore that toothwort's flavor doesn't deter is the caterpillar of the West Virginia white, a Wisconsin State Special Concern butterfly. The butterfly lays eggs on a few species of native toothworts, and the caterpillars must grow up before their short-lived food plants wither away for the summer. In an unfortunate twist, the nonnative, invasive garlic mustard shares chemical cues that attract egg-laying butterflies, but it provides inadequate nutrition for the caterpillars.

What fun to solve mysteries and identify friends along the trail, as living things all race to make the most of spring.

The Changing Woods

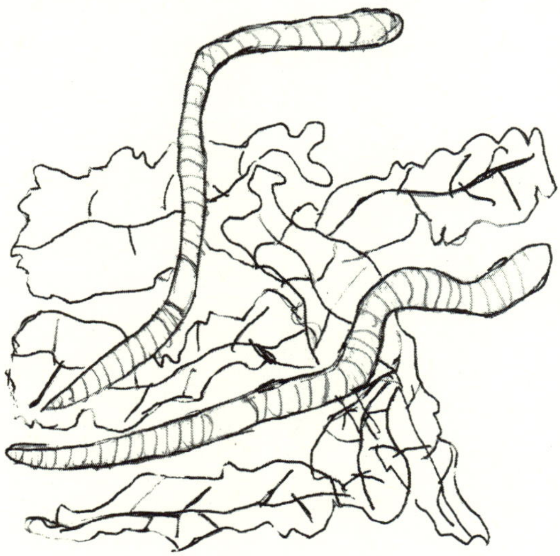

Hael Dums, age 8

The flutelike notes of a hermit thrush wafted through my bedroom window. "Whyyyyy don't you come with me?" he sang in a rising scale. I didn't even open my eyes in the predawn gray, but I did smile to myself. Spring is finally here.

My favorite way to celebrate spring is with a hike up St. Peter's Dome in the Chequamegon-Nicolet National Forest. Gentle slopes, plentiful moisture, deciduous trees, and interesting rocks enrich the soil, which supports an amazing array of wildflowers.

I suspected that I might still be a little early for the majority of blooms, but I had to see what was up. Cut-leaved toothwort, blue cohosh, wood anemone, bloodroot, and wild leeks all poked their little leaves and buds out of the floodplain of Morgan Creek. The pale, fuzzy leaves of wild ginger were beginning to sprout from its rhizomes that snaked along, half-buried in the rich, moist soil.

The wild ginger caught my eye because its rhizomes were more exposed than usual, with a few cloaks of leaves here and there. Crouching down for a photo, I wondered if the foot traffic of people avoiding a mud puddle or some other culprit caused the unveiling. Then I moved a little clump of dead leaves, and the tiny, pinkish point of an earthworm shrank away from the sunlight. Aha!

While European earthworms (brought here with ship ballast or in root balls) are wonderful at breaking down organic matter and mixing the soil in gardens, they are too efficient for the plants in the woods. Northern forests evolved without earthworms, after the glaciers froze their ancestors out. Many plants here need thick, slowly decomposing leaf litter to grow and for their seeds to sprout. In this exposed patch of wild ginger rhizomes, with the continued possibility for nighttime frosts and a lengthening drought, it is easy to see why fallen leaves are important.

Repercussions of the worms' voracious feeding extend throughout the forest. The three-leaved clumps of trilliums unfurling themselves along the trail decline in worm-infested habitats. Other common summer plants, like arching stems of Solomon's seal, thickets of wild sarsaparilla, and even sugar maple seedlings—the next generation of maple syrup—are also at risk. In their place, a lovely, green lawn of Pennsylvania sedge fills in among the trees.

With the lower diversity of plants on the forest floor, ovenbirds (a vociferous, ground-nesting warbler) are much more susceptible to nest predators like squirrels and other birds. One study, conducted in the Chequamegon-Nicolet National Forest and published in 2011, found that ovenbird numbers declined by about 25 percent in worm-infested maple forests versus those without worms. Happily, the same research project also found that three other species of ground-nesting birds—the hermit thrush (who woke me up), the black-and-white warbler, and the veery—are not affected by earthworm invasions.[6]

As I put more space between me and the trailhead, the signs of worms faded quickly. Tiny, white, pants-shaped buds of Dutchman's breeches readied themselves to open next to pink-striped spring beauties who were already attracting bees. Miniature leaves at the tips of chokecherry twigs glowed crimson in the sunlight. I warmed up as the trail steepened.

Spotted salamanders have already returned to a vernal pool under the top of St. Peter's Dome. Their clusters of round eggs—clear with black centers—rest among films of algae, layers of leaves, and tiny, wiggling larvae. I didn't see any signs of worms up that high, which is fortunate. One study looked at the response of salamander populations to earthworm invasions and found that "salamander abundance declined exponentially with decreasing leaf litter volume."[7] Salamanders need lots of leaf litter to house the little critters they eat.

These woods are changing—by the century, the year, and the hour. Some changes—like the graceful unfurling of petals, the delicate emergence of leaves, and the lyrical return of warblers—are welcome. Other changes may be more troubling.

Return of the Ruby

Erin Vaillancourt, age 11

After clicking on a bookmark in my web browser, I anxiously waited for the page to load. A map of the continental United States appeared slowly and the animation started. White, yellow, and orange dots progressively filled the southern tier of states, until a layer of dark red dots finally reached the level of central Wisconsin, just south of Cable.

"Hurry up, they're almost here!" I jokingly admonished my housemate who was scrubbing out our old hummingbird feeder at the sink.

Every spring I watch the migration maps (you can check them out at http://www.learner.org/jnorth/) and eagerly anticipate the arrival of migrating birds. First it's the loons (who have been yodeling on my lake for a few weeks now), and then the ruby-throated hummingbirds.

Another indication of the hummers' impending appearance is the return of yellow-bellied sapsuckers. Sapsuckers' squeaky-toy calls filter through the forest about two weeks before the first hummingbird buzzes in. I spotted my FOY sapsucker several days ago, so it's about time! (FOY is birder speak for first of the year.)

Why is there a connection between the arrival times of sapsuckers and hummingbirds? They have a special relationship. Sapsuckers drill tiny holes in trees. These sapwells provide them with a variety of nourishment. Sweet liquid comes first, followed by sugar crystals as the leaking sap evaporates. Finally, insects who come and feed at the sapwells become protein-packed snacks.

Hummingbirds know this too. The males—who must arrive early to claim a territory—have the challenge of finding food before many flowers are blooming. Cleverly, they will seek out sapsucker trees as a source of tree "nectar" and insects. Since hummers are relatively small, sapsuckers seem to act charitably toward their tiny dinner guests. Freeloaders they are not, though. Male hummingbirds are extremely defensive of their food sources and will chase off a multitude of other birds and critters who might sip more than their fair share of the sap. Thus, both the hummingbirds and the sapsuckers benefit.

Of course, hummingbird feeders can also be an important source of nourishment to the early birds. Hence, our push to get the old feeder out of storage and hung up outside. From my housemate's furious scrubbing, frustrated noises, and questions like "How do I get the black stuff out of that crevice?" it became clear that we were due for a new feeder.

It is important to keep hummingbird feeders sparkling clean and change the water before the black stuff starts to grow. This one had been put away dirty last fall and was now a lost cause. Sun-brittled plastic and broken bee guards finalized the decision.

Later, as I stood paralyzed by indecision in front of the vast selection of hummingbird feeders at the hardware store, something else caught my eye. Big jugs of bright red liquid sat on the bottom shelf, advertising "Ready-to-Use Hummingbird Nectar!"

"No!" I gasped in disgust, startling a customer at the end of the aisle. Red food coloring is a dangerous additive to hummingbird nectar. No, it's not the carcinogenic Red Dye No. 2 banned in 1976. But its replacement, Red Dye No. 40, has been banned in Europe, while simultaneously being approved by the USFDA for consumption in small quantities.

The catch is that hummingbirds don't consume nectar in small quantities. To power their flight muscles (at 53 beats per *second*) and their metabolism (at more than 1,260 heartbeats per minute), hummers eat more than twice their own bodyweight in nectar and insects each day. At this rate, while including a normal proportion of flower nectar, feeder nectar, and insects in their diet, the average hummingbird would ingest 12 times the amount of red dye shown to cause DNA damage in mice. And the mice weigh four times more. While there is no money available to fund a study on the effects of red dye on hummingbirds, it is safe to say that not using it will save you money and probably save their lives.

I shook my head and headed home, bright-red, easy-to-clean feeder in hand. With great anticipation, I poured one-fourth cup of plain white sugar into one cup of water in a Mason jar, screwed the lid on tight, and shook. Soon the crystals dissolved and the water cleared. I filled the feeder, hung it outside my window, and opened my laptop to check the migration map again. A red dot now appeared in Ashland, Wisconsin, a little to my north. They're here!

Just then, I caught a movement out of the corner of my eye. Zoom! He buzzed past the feeder, his ruby throat shimmering in the evening sun. Closing my laptop, I scribbled "FOY hummingbird" in my journal, smiling the whole time.

Bird Muscles and Migration

Vicky Zalatoris

Sometimes it takes a lot of willpower to put on my running shoes and take off down the road after a day at my desk. Today, despite (or maybe because of) the intermittent cloudbursts and beckoned by the ovenbirds' rapid-fire "teacher-teacher-teacher" songs ringing through the trees, I felt it was a necessity. As the cool drops misted my face and the raindrops swallowed sunlight while abandoning themselves to gravity, I felt multiply rewarded for my effort.

"What makes aerobic exercise so powerful is that it's our evolutionary method of generating that spark," according to John Ratey, MD, clinical associate professor of psychiatry at Harvard Medical School, and author of

SPARK: The Revolutionary New Science of Exercise and the Brain. "It lights a fire on every level of your brain."[8]

Then, to add to the magic of the moment, the flutelike song of a wood thrush floated sweetly above the treetops. Henry David Thoreau wrote, "The thrush alone declares the immortal wealth and vigor that is in the forest. Whenever a man hears it, he is young, and Nature is in her spring."[9]

A few more minutes down the road, and a bird flew about five feet in front of my nose. I stopped to look as it landed at eye level. Bird-watching (and listening) is one of my favorite parts of running. Even without binoculars I could tell it was a yellow-rumped warbler, yet another early spring migrant. Soon a whole flock moved through the branches.

The ovenbirds, the thrush, and the other warblers are in the midst of incredible, ancient migrations from their wintering grounds in Central America to their summer nesting territories in the United States and Canada. Their ultramarathon skills put my little four-mile jog to shame, but then again, they are better designed for it.

Birds save energy during flight by having short humerus (upper arm) bones and lengthening their forearms with featherlight structures that are, well, feathers. I can appreciate the difference this makes when I switch from heavy winter boots to my light summer sandals. Every extra ounce my leg has to lift makes each step that much harder! So, light, hollow wing bones, and having the wing muscles located on the chest, concentrates weight close their body and their center of gravity.

By selecting easily digested foods like berries, seeds, and insects, birds were able to shrink their stomachs too. (Just compare a bird gut to the gut of an herbivorous cow.) During their evolution, birds also traded heavy teeth and jaws for lightweight beaks.

With less room needed for their digestive tract, birds have room for more air. Their unique lung system allows birds to route air *through* their lungs instead of in and out like we do. This makes their fresh air exchange incredibly efficient and allows species like bar-headed geese to fly over the top of Mount Everest—without supplemental oxygen!

In the fading sunlight, the flock of warblers flitted effortlessly as I huffed and puffed down the road. I confess, I have a bit of lung envy.

But birds' aerobic capacity isn't simply due to the fact that they take in fresh air more efficiently than we do. The hemoglobin in birds' blood has a high capacity to bind oxygen. Their blood delivers more oxygen per unit than ours. Another chemical, myoglobin, takes oxygen from the blood and delivers it to the cells. This deep red protein is what colors the dark meat on your Thanksgiving turkey. The combination of all these things gives birds an incredibly high VO2 max, or the maximum amount of oxygen the body can use during intense exercise. It's a measure of aerobic capacity that will sound familiar to many athletes.

Not only do I wish I had their lung capacity, I also wish I could eat like a bird. Right before migration, birds feast like crazy and can double their weight in as little as 10 days. And it's healthy for them! The fat gained during this period of hyperphagia is stored in masses under the skin on the abdomen and upper chest. It can fuel nonstop flight for three days and nights. Then the birds are lean again.

While birds have adaptations for athleticism that I can only dream about, we do share one behavioral adaptation: both the birds and I choose to exert ourselves when it is cooler to avoid dehydration in the heat of the day.

"Birds are not likely driven by the logic of what they do," biologist Bernd Heinrich philosophizes in his book *Why We Run*. "Instead, they are motivated by powerful urges. They behave in ways that feel right and pleasurable to them. Feelings of pleasure are a byproduct of evolution that makes healthy organisms do what helps them survive and produce offspring, in the same way that fear makes them shy away from danger."[10]

What makes birds begin their migration, Heinrich says, "is probably not fundamentally different from what motivates me to jog down a country road on a warm sunny day."[11] And, by the end of our respective migrations, both the birds and I feel like singing.

Spring Phenology

Madison Gingras, age 13

Birdsong and sunshine poured through open windows this morning, wafting in on a warm breeze. The distinctive call of a black-throated green warbler was what finally lured me outside. "Zee-zee-zee zoot-zee," he shouted. "This is my territory and don't forget it!" "Welcome home!" I felt like yelling back.

Several other melodies joined his refrain: the trill of a yellow-rumped warbler (a song I'd been hearing for a few weeks already); the "Hey Sweetie" of a black-capped chickadee (I'd been hearing that since January); and the tinkling stream of clear notes bellowed out by a winter wren with incredible breath control and no desire to let anyone get a word in edgewise. A pair of phoebes, and a few other mystery birds, also contributed their verses to the epic ballad of spring.

As I scanned the treetops to catch a glimpse of the feathered minstrels, I noticed with dismay that the maple trees were glowing brightly with tiny, translucent baby leaves among big clusters of dangling flowers. New life is beautiful, but leaf-out makes birding that much harder. I also heaved a sigh at the swarm of freshly hatched black flies vying for my blood. Some signs of spring are not so pleasant.

It's good for the migrating warblers, though, that the insects the birds eat are growing. When the long-distance migrant birds react to changing day length in South America and start their journey north, they have no way of knowing what weather will await them on the next continent. If spring is early up here, they can't get bumped up to an earlier flight. If spring is late, they don't get a text notifying them that their flight has been delayed.

These long-distance migrants tend to arrive like clockwork each spring, with little change in their arrival dates. What variability occurs is mediated by the weather. As the birds head toward their breeding grounds, storms and lack of food will slow them down when spring comes late, and favorable conditions will encourage steady northward progress when spring arrives early.

While the weather this spring has been lovely for a few weeks now, this is the first morning I've heard the black-throated green warbler, who began his journey in Central America. The yellow-rumped warblers (who have been here for a while now) may have only been wintering in the southern United States. The Blackburnian warbler is still on its even longer journey from South America. His arrival date may shift less because it avoids the variability of early spring.

In the meantime, the plants (who can't go south for the winter) are controlled more closely by spring temperatures—although some regulation by day length plays a role in protecting them from blooming during a January thaw.

Spring weather is always variable, but as climate change intensifies, the variability will trend earlier. Short-distance migrants might be able to shift their patterns and stay in sync with plants and insects on their breeding grounds. Researchers have found that the adaptation doesn't happen on the level of the individual, though. Individual birds arrive at the same time year after year. Instead, natural selection favors birds who are programmed to arrive earlier. They have more babies with similar early genes, while the genes of their late-arriving peers go back to the dust.

While this seems a little morbid as I stand surrounded by glowing, singing, flowering, buzzing life in the most vivid season of the year, that adaptation through natural selection is a key to the future of life on Earth.

Winter Wren

Susan Core

The symphony of spring grows richer every day. Stepping out my front door this morning, a blast of birdsong overwhelmed my senses. The translucent, glowing greens and pinks of newborn leaves were muted by low, gray skies, but the warblers' songs lit up my brain with a rainbow of synesthesia.

The upward buzz of the northern parula, who just arrived, sounds blue-gray with yellow and orange highlights. The sweet warble of a Blackburnian warbler sounds blaze orange as it ends in a high pitch I can barely hear. The "I am black and green" song of the black-throated green warbler sounds (what else?) like alternating dashes of black and olive.

Okay, so maybe I don't have synesthesia, maybe it's only that the birds' songs trigger their photos in my mind. But the dreary morning felt decidedly more colorful with the addition of so many perky songs.

These warblers are not known for being quiet. And yet, when the winter wren chimed in with his clear stream of tinkling notes, the other songs faded into a background jumble. This four-inch-long, one-half-ounce, drab-brown bird of the forest floor can sing with 10 times more power than a crowing rooster per unit weight. He's been singing here for weeks now (only needing to migrate from the southern United States), and—although I'm always excited for the new arrivals—his song still gave me a thrill.

I (and the lady wrens) am drawn to him partly by the power of his song, which can last up to eight seconds as he bellows notes while both exhaling and inhaling. And partly I'm intrigued by the mystique of this tiny, well-camouflaged bird who spends his time scurrying through the underbrush with mouse-like hops. The scientific name of the winter wren *(Troglodytes hiemalis)* means winter-loving cave dweller, but their caves are hollows and cavities in dead and fallen trees.

What do they look like? A little brown blur that sometimes congeals into a short, perky, twitching tail. Only through the patience of photographers do I know that they have tiny, white checkers on their brown background, a white eyebrow, and light-brown legs.

My favorite birds are the ones who kept me company during my transient years, when I worked back and forth across the country from California to Maine. I didn't grow up hearing winter wrens in Iowa (house wrens are more common there). I first knowingly heard one in the Boundary Waters of northern Minnesota, which automatically endeared them to my heart. On still mornings their energetic songs would stream out of the swamps and resound across glassy lakes.

When I moved to the redwoods of California, my morning jogging route took me right past the territory of a wren near a tiny stream. They love riparian areas with tangled underbrush. It was exciting to hear a bird I recognized easily in such a drastically different forest. I was also thrilled to hear the familiar voices of song sparrows while I took kids tide-pooling along the ocean.

The song sparrows didn't sound quite like the ones I was used to back east, although their tone and cadence were quite identifiable. It's well known that many birds, like people, have regional variations in their songs. A Minnesota friend who recently headed to West Virginia for a birding trip joked that she'd be hearing warblers with southern accents.

The winter wren song sounds like such a jumble to me that I can't hear the regional variation. The lady wrens can hear it, though. Several years ago a couple of scientists set out to study the variations in several subspecies of winter wrens across the country. Birders had already noted that the wrens in the eastern United States sound more like wrens in England, and the western wrens share more characteristics with their colleagues from Siberia. They are a widely distributed species.

Or are they? Where the eastern and western types meet in eastern British Columbia, they hold neighboring territories. The males will respond to any wren's song with territorial aggression, but the females only mate with males who sing the appropriate song. I've tried to hear the difference, but the nuances are lost on me. Only by looking at sonograms (technological synesthesia!) can I be as discerning as those lady wrens. You can see them for yourself and read an excellent summary of the research on the Cornell Lab of Ornithology's "All About Birds" website.[12]

The discriminating taste of those female wrens seems to be all that's keeping the two types from hybridizing. Scientists call this a "prezygotic barrier to gene flow," because it prevents mating altogether. If that barrier broke down, mating would result in fertile offspring. This seems tenuous, but it is a strong enough divide that in 2010, scientists classified the North American birds as separate species: the Pacific wren and the winter wren. England and Europe also get their own species of winter wren, which, as the only wren outside North America, they just call a wren.

And, seeing as how I love them all and wouldn't last a minute trying to be a discerning lady wren, I will just call him a lovely sign of spring.

Least Flycatcher

Kate Stolp

Birdsong filtered through the window screen even before the sunlight could. The day before, I'd woken up to the syncopated trill with a blisteringly high finale of Blackburnian warblers. Now there was a new voice in the mix. Each day a new wave of migrants sweeps in, contributing their songs to the growing chorus.

However, I'm not sure I'd call the new addition a song exactly, at least in the traditional sense. The sound was a short, emphatic "che-bec" with an insect-like quality and an accented second syllable. It took me two weeks of my first ornithology class to recognize this noise as coming from a bird. I didn't even hear it at first, much less distinguish it from the other call notes and red squirrel chatter going on around me. Once I learned it, though, I heard it everywhere.

Occasionally I see the vocalist too, although that takes more work as the tree leaves emerge at breakneck speed and fill in the lacy network of twigs. Least flycatchers are cute little birds that weigh barely more than four pennies. Like most of their *Empidonax* flycatcher cousins, they have grayish-olive backs, whitish underparts, a white eye ring, and white wing bars.

They can often be spotted as they hawk for insects, which entails flying out from a perch, snatching a winged snack in midair, and fluttering back to the same branch or another one nearby. This describes both the general appearance and the behavior of a whole group of small, drab flycatchers in the *Empidonax* genus. Their songs are the most distinctive things about them. The reason their voices are so distinctive is what fascinates me.

Other "true" songbirds will spend one period of brain development memorizing songs of adults and the next phase trying to match them. This is true of the high-trilling Blackburnian warblers I heard earlier. If a songbird does not grow up with adults of its own species, or can't hear, it does not develop a normal song. But flycatchers do not need to do either: they do not need to hear other adults in order to produce the typical "che-bec" song, and they do not even need to hear themselves to know that they are saying it correctly.

And yet many birders would still consider these rote repetitions the birds' song. This definition is based more on function than structure. A song, at least for many birders, is simply a sound made to establish and defend a territory or to facilitate courtship and mating. With this definition, one can argue that the drumming of ruffed grouse and woodpeckers, booming of prairie chickens, and the wing sounds of a displaying woodcock are all songs. In contrast, many birds also have shorter, simpler calls, that raise an alarm or keep members of a flock in contact.

While my morning walk was filled with least flycatchers "che-bec-ing" every 20 feet, and Blackburnian warblers trilling away, my evening stroll included a soundtrack by the common loon.

Loons are definitely not songbirds, but the males do have a distinctive territorial vocalization, known as the yodel. It's a maniacal up-and-down scream that clearly conveys the serious and stressful business of defending a territory (often at three a.m. when we're trying to sleep). Each loon has his own version. The amazing thing is that when an invading loon is successful in winning a new territory, he will change his yodel. The new composition increases the difference between his yodel and that of the previous, defeated resident, while being careful not to stray too far outside the norms for the neighborhood.

It's easy to be amazed by the many different ways there are to be successful: the loon's ability to change his yodel throughout his lifetime in response to specific events; the warbler's reliance on a distinct period of brain development; and the flycatcher's innate and immobile programming. Each system works for the bird who uses it. Maybe there's a lesson in that.

The Toads Wake Trilling

Brooke Sapper, age 18

Intense, high-pitched trills seemed to follow me home that afternoon. The hum of my bike tires on pavement and the wind in my helmet didn't drown them out. When I pedaled past upland areas, enjoying the warm sunshine, a brief silence would settle in. But as soon as I dipped down past a wetland again, the sound rose up from among the blooming leatherleaf, sprouting cattails, and budding water calla.

The toads were trilling!

Every spring, warming temperatures and longer days trigger those warty brown critters to try their hand at romance. As with warblers, hummingbirds, and many other species, males arrive on the mating grounds well ahead of females to establish their territories. Then the "boys" commence calling. With throats inflated like balloons, they trill for four to twenty seconds at a time.

As with most frogs and toads, females choose a mate based on the length and strength of the male's trill, and also the quality of his territory. Therefore, it is worthwhile to a male frog or toad to expend a lot of energy in calling. While studying gray tree frogs (another loud calling amphibian), scientists discovered that they spend most of the night shouting aerobically at about 60 percent of their maximum output. But when a female is near, they bump it up to near 100 percent for a short time.

To accomplish these athletic feats of song, male frogs and toads have highly developed body-trunk muscles. Packed with mitochondria, the powerhouses of cells, the singing muscles have the capacity for high aerobic metabolism. Frogs and toads call for such a long time that their muscles must switch from burning carbs to burning fats, just like human endurance athletes.

Those muscles are used to drive air over the vocal chords, producing the surprisingly loud calls. Some frogs and toads can be as loud as a lawn mower. Luckily, they have an internal pressure system that keeps their own ear drums from vibrating excessively and therefore prevents hearing loss in the shouter himself.

In contrast, the silent female frogs and toads have much less body-trunk muscle. Their specialization is laying eggs.

Although adult toads are mostly terrestrial, they lay their eggs in water. Their favorite breeding habitats include shallow wetlands, ponds, lakes, and slow-moving streams. Once a lady toad decides on a suitably musical mate and approaches him, the male will climb onto her back and grip around her abdomen with horny pads on his first and second toes. In this posture, called amplexus (used by both frogs and toads), the male can fertilize the strings of eggs externally as she releases them in two rows. A single female can lay 4,000 to 8,000 eggs, connected in a long, spiraling tube of jelly that reaches between 20 and 66 feet in length.

About as big as a blunt pencil tip, each egg is black on top and white on the bottom. This type of camouflage, called countershading, makes the eggs hard to see both from above (looking down at dark water) and from below (looking up at bright sunshine).

Just like the eggs, freshly metamorphosed toadlets and adult toads are well camouflaged in their habitats using a technique called background matching. Toad skin can even change color from yellow to brown to black depending on temperature, humidity, and stress. If a predator finds a toad, the would-be killer gets a mouthful of nastiness. Glands in toads' skin produce a poisonous fluid that is harmful if swallowed or rubbed in the eyes. Toad tadpoles have these same defensive chemicals.

A couple of days after my noisy bike ride, I floated down an equally noisy river in a canoe. In patches along the river banks, the toads' trilling chorus joined with vociferous warblers, orioles, catbirds, and song sparrows to create quite a cacophony of reproductive fever.

Even after I shut myself in a moving car, I couldn't seem to escape the cacophony. Fiddle music played from a CD. The notes whirled and warbled with emotions every bit as powerful as the toads, and quite similar in purpose. I imagined a barn or dance hall full of people, twirling and smiling—and sizing up potential mates—all in step (or out of step as the case may be) to the lively music. Maybe we humans are a bit like the toads when we try our hand at the age-old, springtime music of romance.

Old Turtle

Ben Westlund, age 11

The old turtle scraped at the sand with her naily toes as the kids gathered in a wide circle around her. Kids love dinosaurs, and today the first- and second-graders got a close-up look at a creature who has existed on Earth for more than 40 million years, with direct ancestors much older than dinosaurs.

The students watched quietly and respectfully as the mother slowly finished excavating a depression for her precious cargo at the edge of the boat ramp's asphalt. We commented on her smooth, algae-covered shell and enormous claws on her webbed feet. Once I caught a glimpse between her hind leg and knobby tail and witnessed a smooth, white eggshell sliding into the nest.

The size and age of a female snapper, and the number of eggs she lays each year, are all connected. A mother turtle will lay a clutch of eggs equal to about 7 percent of her body mass each year, and some years she'll lay no eggs at all. This helps to make sure she'll have enough energy to survive the winter, and translates into somewhere between 11 and 87 eggs, with an average of 34 eggs per clutch in northern populations.

Because of this trend, female snapping turtles don't mature until they are eight inches long—big enough support a clutch of about 22 eggs. With a short growing season in the north, that can take 19 years. This big mamma was well over a foot long. How many moons must it have taken her to grow that big?

One reason that snappers grow so slowly is that they are ectotherms, meaning they use their environment to regulate body temperature. In the summer, they sun themselves to warm up. After a long winter, they have to wait until the shallows reach at least 40 degrees in order to become active. Even then, they don't start eating until the water temperature reaches about 60 degrees. This means that in cold northern lakes, snapping turtles may go nine months without eating.

Once they do warm up enough to eat, more than half their diet is vegetation. Snapping turtles are important scavengers and may improve fisheries by eating the slow, bottom-feeding fish (which are generally unpopular among anglers). Although baby ducks do make the occasional tasty snack, they are a much less common part of the snappers' diet than many people think. Over the course of a year, a snapper will only eat its own bodyweight in food. That isn't a recipe for quick growth.

One consolation for their stingy diet may be that snapping turtles rarely become food for something else. The eggs and little guys are vulnerable, of course, but once their carapace (upper shell) reaches three inches long, they have no more natural predators.

Getting there is the tough part.

The eggs we watched being laid have almost no chance of reaching maturity. For one, their location at the edge of a driving surface is pretty risky. But even in a good location, only about 14 percent of clutches hatch each year. Nest predation, temperature variation, and dehydration all are constant dangers.

Temperature is especially important. The embryos won't develop at temperatures cooler than 68 degrees. Above that, interesting things happen. Because turtles evolved before X and Y chromosomes came into the picture, they developed a system that uses temperature to determine the sex of the babies. At 83 degrees, the number of males and females will be equal. Cooler temperatures will produce more males, and higher ones will develop more females.

It is amazing that this lumbering matriarch has persisted against the odds. Then again, turtles are survivors. Fossils of the most primitive turtle put the age of this group at 215 million years old—about 100 million years older than dinosaurs. Turtles survived whatever caused the dinosaurs' great extinction.

Reluctantly the students left the great mother to her important task and began one of their own. With nets and enthusiasm they caught critters in the weedy shallows. Soon one shout rose above the rest: "I caught a turtle!" With its spiky shell barely more than an inch long, this new stage in the life of a snapper—a baby hatched last year—captured their attention as thoroughly as its mother had.

Dragonflies!

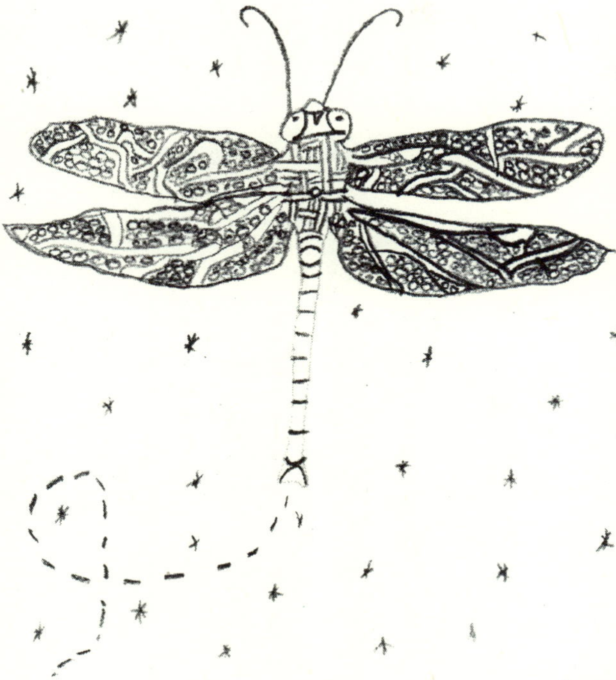

Ella Wald, age 10

Blue skies and a fresh breeze put a spring in my step as I gathered tools—hoe, rake, and shovel—from the shed at the Cable Community Farm. My garden would be fully planted by the end of the afternoon, and I was looking forward to the satisfaction of working the soil, tossing rocks, and nurturing new life.

It was hot, though, in my long pants, long sleeves, and hat. The bugs have been bad at the garden this season, and it's hard to carefully place seeds in a row *and* slap mosquitoes at the same time.

Then, as I carried a flat of squash plants to the gate, I looked up. A cloud of dragonflies darted across the sky with amazing speed and agility. For as far up and as far out as I could see, their tiny silhouettes filled the sky. And then it dawned on me, I hadn't slapped a single mosquito since I got out of my car!

For weeks now, people have been asking, "When are the dragonflies coming?" "Soon," I'd answer, or "I've seen a few by the lake." As beautiful as dragonflies are for their own sake, these queries were really about the mosquitoes. Deep snow, a late winter, and a quick spring had somehow conspired to bring the little bloodsuckers out in terrible numbers. People yearned for the dragonflies to eat the mosquitoes.

Dragonflies are exquisitely adapted aerial predators. Each of a dragonfly's four wings operates independently, powered by its own set of muscles, resulting in fantastic maneuverability. Dragonflies can fly backward and forward, straight up and straight down; hover; accelerate to full speed in a split second; and make hairpin turns. Small antennae seem to have little function except to measure air speed in flight—which can reach 30 miles per hour.

Dragonflies have adapted to see faster too. Together, a dragonfly's eyes and brain can detect movements separated by only one-three-hundredth of a second! Dragonflies also possess an astonishing capacity for selective attention. They can focus on a single mosquito in a swarm, track the moving target, and adjust their path to intercept the prey with a 95 percent success rate. Appropriately, the word *dragon* comes from an ancient Greek word that means "sharp-sighted one."

Three pairs of legs attach to a dragonfly's slanted thorax. This tilts the legs forward and creates a spiny basket used to catch prey on the wing. Small insects can be scooped into the mouth without pause, but larger prey are taken to a perch. There, the dragonfly can masticate the insects with the serrated teeth on their mandibles. These teeth give dragonflies their scientific family name—Odonata—from the Greek word for "toothed one."

While dragonflies can eat their own weight in food every 30 minutes, they aren't reliable mosquito control. Each species of dragonfly completes just one generation per summer, while mosquitoes breed multiple times. In addition, dragonflies eating mosquitoes is like you trying to make popcorn into a satisfying dinner. They prefer bigger prey—even other dragonflies—when available.

Even so, the dragonflies seemed to make a dent in the mosquito population in the garden. So why did they take so long to arrive?

Many dragonfly species synchronize their transition from nymph to adult, which results in a mass emergence. By all metamorphosing at about the same time, dragonflies can ensure that males and females will mature simultaneously and find mates. The cloud of dragonflies will also overwhelm predators and ensure that at least some survive.

Different species are on different schedules, though. While you may notice dragonflies from spring to early fall, you are actually seeing a series of different species. Most adult dragonflies live only a few weeks, although some can live up to a year.

The spring species of dragonflies that zoomed over my head spent last summer as aquatic larvae growing into their final stage before metamorphosis. Then they stopped growing and overwintered as almost-adults, ready to emerge when the water warmed enough.

Now they are fierce aerial predators, patrolling the skies with agility and grace. As I happily tend my garden without long sleeves or bug spray, it's not too much of a stretch to call them angels of mercy.

SUMMER

Bog Buckbean: Nothing Like Its Name

Mimi Crandall

Adults wearing hip waders and wielding nets stepped gingerly into the wetland. An ecologist had briefed them on the importance of aquatic macroinvertebrates for studying water quality, and now it was their job to catch some critters. The public boat landing at Bark Bay Slough State Natural Area near Herbster, Wisconsin, was a gorgeous setting for this adventure.

Bark Bay Slough is a freshwater estuary where the relatively warm, sediment-laden waters of the Bark River meet the clear, icy waters of Lake Superior. The lagoon is mostly separated from the lake by a sandy baymouth bar, making it a peaceful place to study water quality. Within this unique physical setting are equally delightful plants.

Mats of floating vegetation and peat soils creep in from all sides of the lagoon and support both bog and fen communities.

The students in hip waders were participants in a Wisconsin Master Naturalist Volunteer Training Course. For an entire week they were assigned to explore beautiful places with natural resource professionals and learn a little bit about everything. Some folks take the course purely as personal edification, but most students are planning to apply their newfound knowledge to volunteer activities or professional naturalist experiences.

My job, besides coordinating logistics and piping in with cool facts, was to take photos. So, as a couple of students waded across a knee-deep channel to sample insects west of the landing, I followed behind with my camera held high. After pushing through the cattails and finding stable footing, I looked up.

The mat of fen vegetation spread out before me like a magic carpet floating on water instead of air. From this carpet rose a thick bouquet of flowers. Clusters of delicate pink blossoms drooped from the graceful, bowing stems of bog rosemary. Threadlike arcs of fen sedge filled in between. Rising above them were the frilly white flowers of bog buckbean. Its beauty was the crowning jewel.

Despite its slightly hokey and incorrect name, bog buckbean is one of the most magnificent members of a fen community. Bogs and fens are often confused, so that part of the name is no surprise. Bogs are peaty wetlands that receive all or most of their water from precipitation rather than from runoff, groundwater, or streams. They are characterized by acidic soils with few nutrients. Fens are also peaty wetlands, but they receive input from groundwater, runoff, or other sources, which decreases their acidity and increases nutrients. The water of this peatland in the Bark Bay Slough meets and mixes with water from the estuary, which makes the bog buckbean's habitat a fen.

The "buck" part of the flower's name may be a corruption of the name Puck, who was a clever, mischievous elf in old English folklore. According to one account, the children of Devonshire, England, used to address Puck to ask for protection before traveling through the forest after dark.

How the character and the plant came together is still a mystery to me, but this does bring up the fact that bog buckbean is found in wetlands around the globe! From North America to northern Asia, Europe, and Japan, this beautiful flower remains the same species. It's a small world. Also unusual is that *Menyanthes trifoliata* is the only flower in its genus.

One alternative name of bog buckbean is bog hop, which refers to the English use of its bitter leaves to flavor beer. Those leaves are divided into three oval leaflets, and they must have reminded someone of the shape of broad beans from their garden because the "bean" in this plant's name refers to their shape.

It's the magnificent flowers, though, that steal the show. The tall, cone-shaped inflorescence holds many feathery white, starlike blossoms. The feathery petals aren't a simple vanity. Pollination is carried out by bumble bees and other big bees. Small insects aren't effective pollinators, so lacy obstacles on the petals protect the flower's valuable nectar from freeloaders.

The flowers also make another investment in cross-pollination: some bog buckbean plants have a long pistil (female flower part) and short stamens (pollen-bearing flower parts), and some plants have the opposite combination. If a bee gets dusted with pollen from a long stamen, that pollen won't rub off on another flower with a long stamen. Instead, the pollen will be at the perfect height to cross-pollinate a buckbean flower with a long pistil.

Looking back at my photos, all the buckbean flowers in this fen seem to have five short, violet-colored stamens and a tall, yellow pistil in the center. Buckbean spreads more through rhizomes than seeds, so it's quite possible that all these flowers are part of the same wandering clone, connected by a network of horizontal stems.

After photographing the flowers and students, I bounced a few times on the floating mat of the fen and wandered back to the group at the landing. Naturalists with nets had caught some neat little insects. After lunch we paddled out to the baymouth bar and walked across to Lake Superior.

Geology, ecology, botany, biology, and people: all had come together for a special day at a special place.

The Charm of Sundew

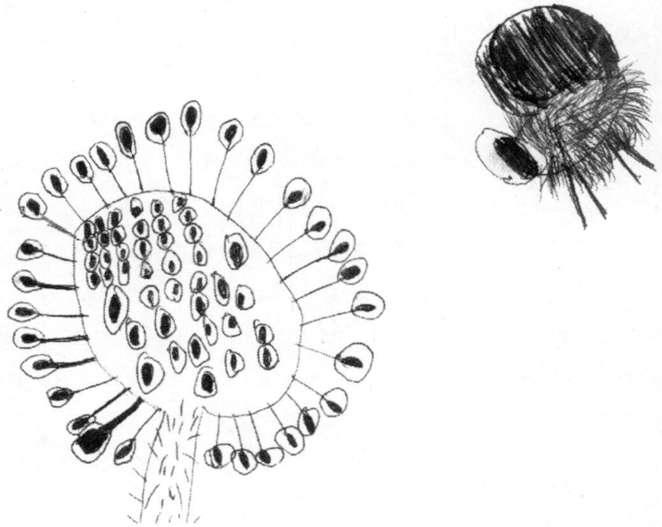

Cian Braun, age 8

A pungent, earthy, vibrant aroma seemed to squirt out of the bog with every sinking footstep. Even through the soles of my muck boots I could enjoy the varying textures of this living carpet. Soggy sphagnum moss offered almost no resistance, while the skeletons of tough twigs buried within the moss crackled, bent, and snapped. At some point in each footfall, the cushion of living and dead plants pushed back toward me just enough so that I didn't fall through.

Bogs are a unique, almost alien landscape with a charm all their own. Peculiar plants, few trees, and a wonderful, squelchy, squashy, shaky, shivery, sucking substrate can turn adults back into giggly, wiggly kids. These 22

Wisconsin Master Naturalist Volunteer students were no different as they spread out to explore the secret bog at the Forest Lodge Nature Trail.

Squatting down next to a domed hummock, I peered into a miniature jungle. Delicate vines of bog cranberry held up tiny pink flowers with swept-back petals. The wiry lattice of leatherleaf twigs and bog laurel's leaning stems with deep green leaves all poked their heads out of the mat of sphagnum moss like drowning rats. Sphagnum moss may be the bog's bully—pulling itself up on the stems of others and using woody plants like scaffolding—but it is also one of the major architects of this strange habitat.

Sphagnum's primitive leaves are like tiny sponges, and they can absorb up to 26 times their own weight in water. This power can raise the water table. Such a saturated environment is low in the oxygen that microbes need to accomplish decomposition, so nutrients stay locked up in dead vegetation. To deal with the lack of available minerals, sphagnum shoots hydrogen ions into its environment, dislodging scarce nutrients for its own use. This acidifies the bog, which further limits what plants can grow there.

Life is not easily discouraged, though, and a special suite of plants thrives in sphagnum's world.

Peering over the side of the hummock, a sparkle caught my eye. Looking closer I discovered a cluster of fuchsia-colored hairy spoons only an inch or two high. "Sundew!" I squealed, pulling over the nearest student to exclaim about it with me.

Round-leaf sundews are beautiful little plants with a circumboreal distribution that spans the northern parts of the globe. They also have a penchant for animal flesh. Well, a taste for insects and spiders at least.

Sundews love the sunny, moist habitat created by sphagnum, but they still need to eat. Nitrogen is a limiting nutrient in bogs. Some plants get their nitrogen through complex root systems and relationships with fungi, while others conserve it with the austerity of evergreen leaves. Sundews catch bugs.

The sparkle that caught my eye was actually a drop of "dew" clinging to the tip of a tiny tentacle. The bowls of sundew's spoon-shaped leaves are dotted with these tentacles. Long ones hold a drop of sweet, sticky nectar that attracts and entraps prey. Once an unsuspecting insect has been mired in this mucilage, the leaf curls inward. This response to touch is known by the comical term *thigmonasty*. From the bug's perspective that translates to "thing most nasty."

Death usually occurs within 15 minutes as the prey succumbs to exhaustion or is suffocated by the goo. Shorter tentacles with drops of digestive enzymes at their tips make contact with the prey and reduce it to nutrient-rich soup, which is absorbed through the leaf surfaces. The valuable nitrogen is used to make chlorophyll, enzymes, proteins, and seeds.

The seeds are produced in a delicate white-petaled flower held aloft on a long stalk. But how, you might ask, does the insectivorous sundew avoid eating its own would-be pollinators? In nature's wisdom a completely different set of bugs is attracted to the flower from those who are lured in by the saccharine dew.

The group of students was lured in too, a few at a time, to look at the impressive patch of sparkling leaves. As we headed back to solid ground for our closing discussion, I plucked a little cluster of sundew leaves to bring in for a closer look. At first I was annoyed when the tentacles stuck stubbornly to my finger. Then I chuckled at my own surprise. It seems that Master Naturalists are no more immune than bugs to the cloying charms of sundew.

Cicadas

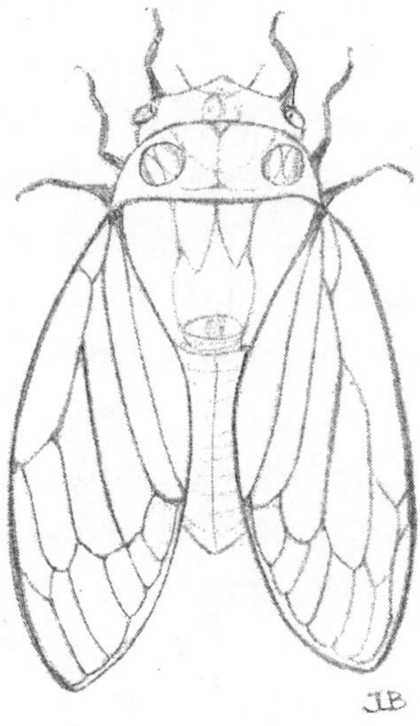

Janet Barthel

I could hear murmured gasps of surprise and appreciation as the Wisconsin Master Naturalist Volunteer students stepped out of the car caravan and into the Moquah Barrens. Rolling hills carpeted in grass and punctuated by pines stretched as far as the eye could see. Vistas aren't common in Wisconsin. Our Northwoods landscape is dominated by dense trees. In contrast, the deep sand deposits left by glaciers on the Bayfield Peninsula provide the substrate for a grassland ecosystem reminiscent of the Serengeti.

Matt Bushman, a botanist for the Chequamegon-Nicolet National Forest, gathered the group to start talking about plants. While he held up sand cherry, smooth aster, and Pennsylvania sedge, the heavy clouds parted and tentative sunshine brightened across the savanna. As the light strengthened, a high buzzing became noticeable and then intensified. The crescendo was so perfectly synchronized that it seemed the sunshine itself came with sound effects.

From behind my head I heard one particularly loud buzz. Not really expecting to see the source, I peeked over my shoulder. To my surprise I easily spotted the noisy cicada as it perched in a small jack pine tree.

Cicadas are true bugs with wide-set eyes, short antennae, and big, clear front wings. They make noise not by rubbing together hind leg and forewing as grasshoppers and crickets do, but by rapidly vibrating a portion of their exoskeleton. The drumlike tymbals are a pair of corrugated membranes on the sides of the cicada's abdomen. As the tymbals vibrate, the male cicada's hollow abdomen acts as a resonance chamber and amplifies the sound.

Before long the sun disappeared above a passing downpour. We jumped into vehicles and bounced along a sandy road to our next site. The rain abated as we gathered to admire a fresh burn. This spring the Forest Service had ignited a prescribed fire in this management block. The ground was still black with soot and charcoal, but already the bracken ferns had erupted in a neon green ruff across the hills.

Sunlight and buzzing filled the air. The class hoofed it uphill, picking their way through the shin-high forest of bracken. Something caught my eye against the black ground. An adult cicada crouched silently. Its huge, clear wings folded neatly over its dark back like a clear plastic poncho. Through the membranes I could see delicate orange bands on its abdomen, orange legs, and more orange markings behind its head. Wide-set, dark eyes fronted its alien head.

Nearby, still clinging to a charred stick, I found an empty shell. This wasn't the body of a dead cicada, rather it was the discarded exoskeleton of a young cicada who had emerged and flown off as an adult. The exuviae, as scientists call them, were everywhere: tumbled on the ground, hooked to vegetation, blown into drifts against the bases of stumps.

As I looked down I began to see their tunnels too—inky black round holes, as if we'd all just poked our fingers in the soil.

Cicadas have a fascinating life cycle. After seeking out a loud male and mating, the female cicada slices open a twig and lays her eggs. Upon hatching, the nymphs drop to the ground and use their strong front legs to burrow in. The nymphs will spend most of their lives about eight feet down, sucking up the sap of plants.

The number of years that a cicada nymph spends underground varies by species. Two to five years is common. The most famous cicadas, though, are the thirteen-year and seventeen-year periodical cicadas of eastern North America. They emerge all at once in tremendous numbers. Their genus, appropriately, is

Magicicada.

Magic indeed. Eyes on the ground, we soon found our own enchanted fairy. Perched on the empty exuviae—which was still hooked to a stick by tiny claws—sat a cross between a pixie and an alien. Huge, clear wings were threaded with glowing white veins. The plump sea-foam green body was etched with pale yellow and orange highlights. Staring straight at us were a pair of false eyes with the vertical pupils of a cat.

Once a cicada nymph crawls out of its hole, it climbs partway up a nearby plant, splits its exoskeleton between the shoulders, back dives free, and then slowly hardens off and darkens in color. With shouts of excitement, we discovered a cicada in mid back dive, its wings still folded like baby leaves.

Magic as these were, they weren't the 17-year cicadas. Those adults have red-orange eyes. A quick internet search brought up the tentative identification of *Okanagana balli*, the short-grass prairie cicada, a species with a shorter life cycle.

More raindrops finally snapped us out of the cicada's enchantment. As we stood up to stretch cramping legs and crooked backs, the vivid green hills and shrouding mists of the Moquah Barrens renewed our sense of awe. Master Naturalist students study everything from landscape-scale geology to the ephemeral lives of insects in their weeklong course. The breadth of topics can be daunting, but words from a friend help to create enough space in my heart for them all. He said, "The world is beautiful at all scales."

A Clay-Colored Sparrow Returns

Kate Stolp

A light rain was starting to taper off as we crashed through the brush. The spicy fragrance of crushed sweetfern tickled our noses pleasantly. Birds sang all around us. Pushing through the clumps of head-high, stump-sprouting scrub oak and red maple, we stepped carefully over the sooty remains of trees. Coming up on a long, narrow open area, a tall metal pole appeared out of nowhere—like finding the lamppost in Narnia.

With a little squinting, we soon noticed some filmy, black netting strung from the pole, stretched 30 feet through the open area, and hitched to a matching pole. A few raindrops sparkled on the web. Here, our small group of Wisconsin Master Naturalist students gathered around

Jim Bryce, a retired biologist with the USDA Forest Service and National Park Service and longtime bird bander.

Jim explained how the mist net functions. The fine black threads are nearly invisible, and they are strung to form long, floppy pockets. When an unsuspecting bird flies into the net, it drops into a pocket and gets tangled. Then a bander can carefully extricate the bird for processing.

As he talked we noticed a flutter at the far end of the net. "Looks like we got one," observed Jim, as he pushed toward it through the brush. And then, with considerably more excitement in his voice, "It already has a band! This must be the same clay-colored sparrow we caught last year!"

Returns—defined as recatching a previously banded bird at least three months later—are rare in the banding world, but they provide a wealth of information. From these recaptured birds scientists have learned about the incredible 24,000-mile round-trip migration that Arctic terns complete each year, the tropical wintering habitat of Bicknell's thrushes, and the maximum known age for a wild bald eagle (32 years, 10 months). However, only about 1 percent of small birds are ever caught again.

"This is the perfect habitat for clay-colored sparrows," Jim explained as he gently untangled the banded sparrow's feet, feathers, and beak from the net. The Forest Service has been managing this area as open brushland by using fire to keep out the trees and allow the growth of dense, shrubby cover that many species need. That's why we were stepping over burned logs and pushing through thickets of fire-adapted plants on the way here.

Clay-colored sparrows need this brushy habitat. Females build their nests in the lower branches of impenetrable thickets, preferring a spot with as little light penetration as possible for protection from predators. A low nest is good for the fledging chicks too, since they hop to the ground before they can fly and hide in dense shrubs nearby while their parents feed them for a few more days. Unlike most other species, adult clay-colored sparrows forage outside their nesting territory throughout the whole process. This leaves more food available near the nest for the flightless chicks.

Mostly due to a loss of their preferred brushy habitat, clay-colored sparrows have been in a slow decline over the past 40 years. That is one reason why the Forest Service's management plan here in the Moquah Barrens on the Bayfield Peninsula is so important. Our returning sparrow is proof that the Forest Service is doing a good job. Jim's annual bird banding here will help to quantify their success.

When we finally got the sparrow extricated from the net and back to the processing table, we confirmed that he was in fact the same bird—band 53262—my Master Naturalist course caught last year in that same place. Male clay-colored sparrows return to the same territory year after year, but pairs do not mate for life. Instead, each year during migration the females find a new mate and follow him home.

The government will be following him now too. At the end of each year, Jim reports all his banding data to the US Geological Survey Bird Banding Laboratory. They'll make note of this return. The Forest Service will be interested in that information, since it helps to validate their management plan. If our little friend is caught a third time, the one who captures him can report it to the laboratory, and all interested parties will gain even more information.

Jim spent a few minutes with our bird, measuring his wing length, tail length, and weight. Once the data was collected, a student placed our little brown-and-white streaked friend gently on the ground. We held our collective breath for a second as he got his bearings, and then with a flutter he was gone.

As we turned back to the table to process more birds, we heard the characteristic "bzz-bzz-bzz" song of a clay-colored sparrow claiming this beautiful classroom as home.

Hooked on the River

Connor Abric, age 15

It didn't take long to pack their Duluth packs. Six of the seven boys had been on a canoe trip with me before; the newbie was a younger brother. Sleeping bags, sleeping pads, headlamps, and mess kits from the National Park Service went into the garbage bags inside their packs, along with their clothes, toothbrushes, and extra shoes. Within half an hour seven boys, two leaders, and two volunteer drivers were on our way to a Namekagon River landing. (Just to be clear, girls are invited too, but sadly, none signed up.)

Many hands make light work. Unloading the boats went quickly. I'm often impressed with these local kids—they jump right in to help and don't complain. Before launching the canoes we slathered on sunscreen and fueled up with gorp (a.k.a. trail mix). "Two hands," I reminded them as I poured peanuts, raisins, and M&Ms from the bag. To drop an M&M is a major fail.

Soon we were floating. A pair of great blue herons

ascended like dinosaurs from the reedy shallows around the first bend. Kingfishers swooped overhead, and cedar waxwings gave their high, thin whistles from the shrubs on shore. The boys didn't notice the birdsong, though, as they focused on navigating through quick water and avoiding rocks. One canoe got distracted by some fishing line tangled in the alders, and the boys managed to free quite a bit before moving on. They've been told how dangerous lost line can be to wildlife, and a use for the line was already swimming in their heads.

A light drizzle gave way to tentative sunshine as we landed at our first night's campsite. After we gathered around the picnic table, David, an intern with the St. Croix National Scenic Riverway and my coleader, asked, "Does anyone know where the closest national park is?" "Right here!" several boys replied. These boys are unique: they've heard this spiel before. Many local residents drive by the brown arrowhead road signs every day and never stop to appreciate the treasure they signify. The Namekagon River is a tributary of the St. Croix River, and therefore is part of the St. Croix National Scenic Riverway, which is a unit of the National Park Service (NPS).

Both the Cable Natural History Museum and the NPS want to get kids on the river. So, the NPS and the Museum have developed a marvelous partnership in the past few years. I provide the trip guiding experience, camp food, and participants, and the NPS provides camping gear, a second adult to come along, a ranger for an evening program, and logistical support. In 2014, funding was provided through the St. Croix River Association, which was awarded an America's Best Idea grant from the National Park Foundation. As I tell our NPS partners: just like the algae and the fungus in lichens, we couldn't do it without each other.

The boys are the ones who really benefit, though. We covered our miles quickly on the first day, and intermittent drizzle dampened their enthusiasm for swimming. So what could they do around the campsite? Before long one kid was making a fishhook by filing down the metal end of bungee cord found near the fire ring, another was whittling a fishing pole from a stick, and a third was untangling the fishing line found earlier.

Not a single boy pulled out an electronic device or even mentioned wanting one. Two Frisbees (one of mine and one found in the weeds on the river) were sailing among the tents. A few other boys were practicing intently with David's *kendama*, an extremely addictive wooden skill toy that originated in Japan.

No one was bored. Everyone was safe. I sat back and let them be boys in the woods.

Soon the fishing pole was outfitted with a wooden bobber, baited with a raisin snitched out of the gorp, and ready for testing. The whole troop followed Gavin down to the landing and watched as he swung the hook and line into the current. Evening light reflected off the trees, and bugs skittered along the surface. The onlookers soon dispersed, but Gavin stood quietly—our most talkative boy sliding easily into the fisherman's meditation.

After several minutes without any nibbles, Gavin decided to try the other canoe landing—a backwater area filled with water lilies and muck. As he waded in wearing sandals, I heard him mutter, "I hope I get a leech on my foot so I can use it for bait."

At some point we all drifted back around the picnic table to watch the *kendama* practice. Then out of the blue, Gavin exclaimed, "I've got a leech!" with not a hint of disgust or fear in his voice. Nearby, Grant, who had whittled the fishing pole, replied, "Sweet! Now we have bait!" with the same untainted joy.

The boys never caught a fish with their makeshift pole and leech bait, but I know they caught the spirit of the river: adventure, resourcefulness, stewardship, and beauty. Over the next two days we paddled rapids, cooked on a campfire started with flint and steel, learned about voyageur history, picked other people's trash out of a fire grate, admired bald eagles and osprey soaring above, and paddled 15 miles in a single morning.

Around the campfire I asked the kids what they wanted in a trip next year. "More fishing!" exclaimed one. "Can we go longer?" suggested another. Others echoed those thoughts. Obviously, we're all hooked on the river.

Boys on the River: Imitators and Attractors

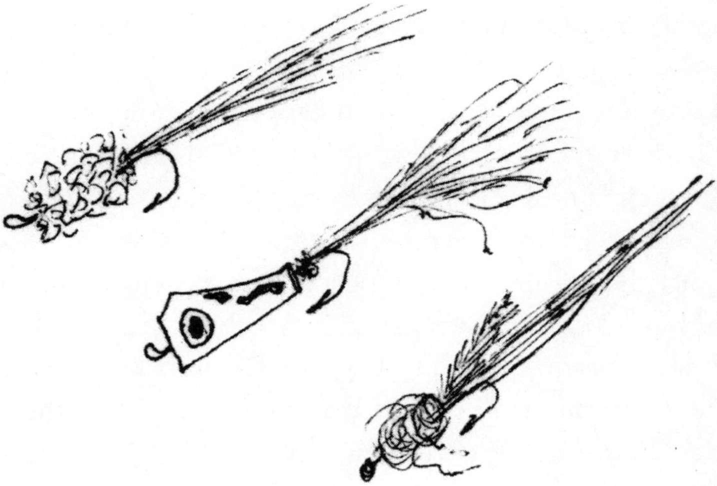

Ranger Jeff Butler

The first 10 miles went by quickly. Once their paddles hit the water, the six teenage boys on the 2015 Paddle the Namekagon canoe trip didn't want to stop. They loved looping around wide, lazy corners, bouncing through short sections of mild rapids, and getting the hang of steering. They also learned how to duck and how to extricate the boat from tree branches when the steering didn't work. My coleader (a National Park Service intern) and I gave them paddling tips, encouragement, and a lunch break, but mostly we hung back and let them be boys on a river.

Arrival at the campsite meant learning how to set up

new tents, laying out sleeping bags, and then an eagerly anticipated swim.

With the boys cooled off and tired out (for a few minutes anyway), our guest instructor took over. John Kudlas is a longtime outdoorsman and Cable Natural History Museum volunteer. As a retired high school biology teacher, he is a wealth of knowledge and skills. Soon the boys scattered away on a scavenger hunt that sent them looking for aquatic plants, identifying forest trees, and even collecting macroinvertebrates—the insects that spend all or part of their lives in the river.

With practiced efficiency, and his usual humor, John made sure the boys knew how important those invertebrates are. Both the aquatic nymph stages and the flighted adults of things like stoneflies and mayflies are water quality indicators, as well as good fish food, he explained. At the mention of fish, I saw the boys perk up their ears. As a surprise treat, John whipped out fish hooks, homemade hook vises, spools of thread, feathers, and fur, and soon the aspiring anglers were creating artificial flies for fly fishing.

This was delicate work, and the boys' concentration was palpable. I noticed the tip of at least one tongue bitten in concentration. The long, brown nymphs each boy soon created were "imitators," or artificial flies that seek to deceive fish by being a lifelike imitation of their insect prey. The boys were in good company. Humans have been creating imitators since at least AD 200, and the term *artificial fly* was made popular by Izaak Walton's 1653 book, *The Compleat Angler (sic)*.

By the mid to late 1800s fly fishermen had expanded from tying flies that imitate prey to tying "attractors"—colorful, abstract flies that don't look like prey but somehow incite aggression and attract strikes from fish.

Our boys followed this same progression in their fly tying too.

After an easy morning paddle, we met Ranger Jeff at a large landing site. Since the Namekagon River is a National Wild and Scenic River—a unit of the National Park Service—we got incredible support from a host of park rangers along our journey, and also some funding from the St. Croix River Association.

Ranger Jeff provided us with some real bling. He had a picnic table at the grassy landing set up and waiting for us. Here were professional hook vises for each student as well as brightly colored foam sheets, vivid pink and naturally colored feathers, sparkly tinsel, and "scope for the imagination."[1] Ranger Jeff has a passion for teaching the next generation about fishing, and it showed in his enthusiasm, preparation, and carefully thought-out teaching progression.

As Ranger Jeff walked the kids through the fly-tying steps, I could see both their eagerness to do things right for Jeff and their confusion over how this odd hodgepodge of materials would become an artificial fly. Cutting foam, wrapping thread, folding, wrapping again, adding feathers, and wrapping some more—both the method and a form gradually took shape. Soon we had a swarm of flashy artificial flies with polka dots, eyes, tails, tufts, and even a comical unibrow.

Ranger Jeff then gave the boys some free time to stretch their legs and get out their wiggles. He also put out a book of fly-tying patterns, opened up a case of professionally tied flies, and gave the boys the option of tying a second fly. After about 10 minutes of Frisbee, they were all back at the table, heads bent to the task.

With the basic techniques mastered, the boys had soon created a second swarm of flies, this time with multicolored mantles, artfully layered feathers, and gracefully flowing tails. Though not imitators of life, these works of art looked like they could attract a crowd, and hopefully a fish.

Then the time came for the real test. Ranger Jeff rigged each of the boys up with a fly rod and an artificial fly—hook and all. We waded into the shallow water at the river's edge and began to cast. Back and forth the lines arched gracefully (mostly) over their heads and out into the current. Flies landed briefly on the surface, dancing in the ripples. Once again the boys' concentration was palpable. Cool water rushed around their knees. Flowing sand nestled around their feet. The jokes paused. Time slowed.

Imitators: Ranger Jeff stood in the middle of the row of young anglers casting again and again. I noticed as the boys glanced sideways at him from time to time, then turned back to their own rods in an effort to emulate his graceful motions.

Attractors: The river is its own lure—its peace, its beauty, its constant motion; the promise of challenge, adventure, and fun. And the possibility that if you tie the perfect fly and cast the right way, you'll feel the strike of a fish on your line.

In the preface of *The Compleat Angler*, Izaak Walton wrote, "No man is born an Artist nor an Angler." That may be true, but I think this trip inspired a few of each.

Monarch Chrysalids

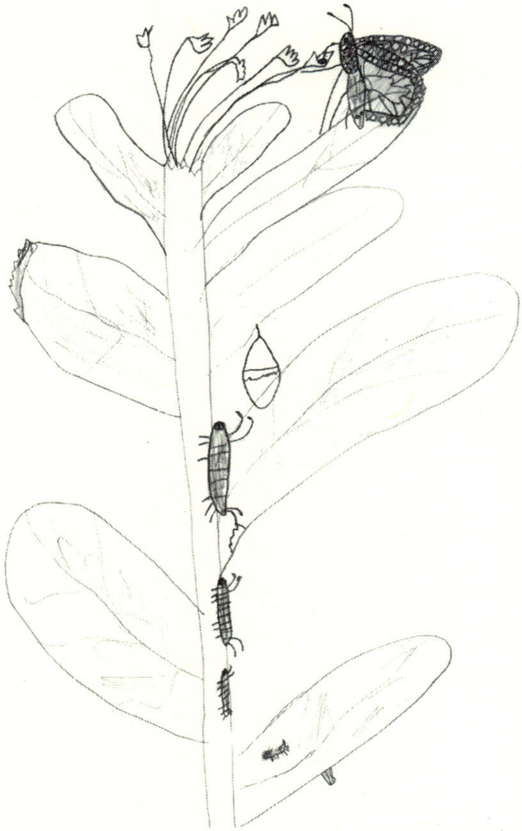

Eve Depew, age 9

A week had gone by and I hadn't had a chance to check on the monarch caterpillars I'd found on milkweed growing near the Cable Natural History Museum's front door. So after I'd waved good-bye to the last fly-tying, canoe-paddling, gorp-eating boy, I deserted the pile of dirty camping dishes in the sink and rushed out to check on my caterpillar crew.

The milkweed looked like it had seen better days. Many leaves were completely missing, while others were chewed down to a yellowing stub. While the evidence of their voracious eating was all around, a quick glance revealed no caterpillars. I searched harder and harder: checking under each leaf, peering under the pendulous flowers, until I satisfied myself that the caterpillars were nowhere to be found.

A little research confirmed what I suspected. Once a monarch caterpillar has reached its final instar and stops eating, it enters a wandering stage and travels some distance from its host plant to find a safer, less visible place to pupate. Somewhere in the sweetfern thicket on the Museum's berm may hang four delicate pale-green chrysalids. (Side note: If you're wondering about my spelling, both *chrysalis* and *chrysalid* are correct terms for the pupa, and *chrysalid* is easier to make plural.)

Since I had hoped to have a front row seat during the caterpillars' transformation into adults, I was more than a little disappointed. The blinking orange light of my voice mail soon changed that.

Bob Olson, one of my readers, had to tell me about his granddaughter, Eve Depew. This bright-eyed seven-year-old was raising more than 50 monarch butterflies on her back deck in Hayward, Wisconsin. She (with the help of her supportive parents) plans to save the monarchs from extinction by releasing her butterflies into the wild. I followed up on Bob's tip and soon had a visit scheduled to

Eve's backyard butterfly conservatory.

Later that week as I chatted with Eve and her mother, Brianne, near the wood and netting butterfly nursery that Eve's dad had built, we kept one eye on the action. A neat line of chrysalids dangled from the sunny side of the frame, interspersed with crumpled, empty shells and their orange-winged owners. (It was too cold to release the seven new butterflies today.)

Hanging in one corner, a brightly striped caterpillar stuck out like a sore thumb. "He's in the J," spouted Eve. "I hope he changes soon!" Looking closer she noted how the white stripes were looking greenish, and she amended her exclamation: "I think he's almost ready!"

While this caterpillar still looked more like a larva than a butterfly, some important changes had already taken place. During the wandering stage (when Eve must rescue wayward caterpillars off the deck furniture), veins developed that will supply his wings with fluid. As the caterpillar pupates, each wing is shaped and readied for adulthood.

During a pause in the conversation we looked over to see that the caterpillar's J was relaxing. "He's ready!" announced Eve, swinging her blonde braids over her shoulder.

Sure enough, the skin near the back of his head had begun to split, and a small triangle of green peeked through. Almost imperceptibly this triangle grew as his exoskeleton sloughed off and crumpled into a wad near the silk button that attached it to the wooden frame. It is easy to think of a caterpillar spinning a cloak around itself, but the delicate green shell of the chrysalid is actually what the caterpillar reveals during its last molt.

I stuck my nose up to the screen for a front-row seat while mother and daughter giggled together, narrating the action and imitating the energetic wiggling and twisting the caterpillar must do to kick off the last of its old clothes without falling. This is no easy feat, as the monarch must pull a small, sticklike cremaster out of its exoskeleton and poke it into the silk button to secure its attachment for the long haul. This is a risky time for the monarch, and Eve told me sadly of one pupa that failed this step, its life ending in a pile of mush on the ground.

The newly formed pupa looked lopsided to me, and upside down. Eve assured me that this was normal. In a few hours the chrysalid's exoskeleton would shape up and harden off.

Throughout my stay both mother and daughter impressed me with their detailed knowledge of the life cycle. Only when I asked about the mysterious changes that happen inside the chrysalid did Eve furrow her brow and say, "I just don't know." Scientists are only beginning to figure out the details themselves.

While the pupa is soft and vulnerable as it forms, it isn't quite the soup of cells that some people imagine. At no time do all the body parts break down, although the chewing, crawling muscles of the caterpillar are reduced to liquid and reformed into the flight muscles of a butterfly. During the caterpillar's final instar, clusters of cells called imaginal disks start to grow rapidly. Besides the wings, they have already formed a beating heart, a respiratory system, antennae, legs, and a proboscis. Those parts and others need more time to develop fully, and old caterpillar parts need to be digested and reabsorbed.

As I sat transfixed by the show, Eve told me how she gathers fresh milkweed (complete with hitchhiking eggs and caterpillars) from a neighbor's field every day and cleans the copious caterpillar poop out of the enclosure. "She spends hours out here," her mother said with a proud smile. "I think she's going to grow up to be a scientist." In my opinion this little caterpillar is already well on her way.

Higher Plants

Vivianne Hanke

The view was spectacular. Hot and sweaty, my friend and I emerged from a humid forest into brilliant sunlight on top of the world. Our toes crept toward the edge of lichen-crusted rocks and our souls soared out over the deep blue waters of Lake Superior.

Hiking to the Top of the Giant, in Sleeping Giant Provincial Park in Ontario, Canada, has been on my adventure list for a while. Despite the July heat, the month of record rainfall collecting in mud puddles, and a bumper crop of bugs, it was worth the wait. Forested hills spread their welcoming embrace, craggy cliffs rose wildly from the blanket of green, and Lake Superior sparkled with its special magic.

After inhaling the grandeur for several moments I turned my focus downward again, perhaps seeking relief from such strenuous awe.

The world is beautiful at all scales. There at my toes were the exquisite flowers of three-toothed cinquefoil. On each small blossom five white petals threw themselves back like outspread arms embracing the sky, and a crowd of delicate stamens and pistils reached eagerly toward the sun. Shiny green leaves spread out in a low mat. Each leaf was palmately divided into three lobes, and each lobe was tipped by three pointed teeth.

These are the mountain men of the plant world. Three-toothed cinquefoil can't survive on good soils where numerous neighbors crowd close. Instead, they claim space where others can't survive and dig their toes into craggy bedrock. Here they eke out a living on dust, rain, solitude, and beauty. Woody, creeping stems cling to the available footholds; thick, leathery leaves defend their precious water; and red pigments protect their winter foliage from harsh sun.

They've come a long way.

Just over 1.1 billion years ago, two separate events initiated this combination of rock and flower. These craggy cliffs where cinquefoil clings solidified from magma 1.1 billion years ago during the Midcontinent Rift, the event that initiated the Lake Superior basin. The continent tried to rip itself in two. Farther south, along Lake Superior's North Shore in Minnesota, lava erupted onto the surface. Up here in Canada molten rock squeezed between layers of older sedimentary rocks. The magma cooled, the rocks tilted a bit, and then the softer sedimentary rocks began to erode away. This resulted in the spectacular topography and our steep hike.

The second event occurred in an ancient sea where the first cells of life swam in a primordial soup. One eukaryotic cell, now known as an archaeplastida, engulfed a free-living, photosynthetic cyanobacteria. The two single-celled organisms began living in symbiosis. Some scientists say that the eukaryote enslaved the cyanobacteria, forcing it to produce food for them both. Others acknowledge that the surrounding cell would have given the cyanobacteria a measure of protection, plus access to the sunlight, water, and carbon dioxide it needed to photosynthesize.

Eventually the two different cells began to coordinate their division (and hence their reproduction) and even trade genes. Eventually the cyanobacteria was no longer a free-living endosymbiont. It was a chloroplast, just one of the organelles in the eukaryotic cell.

Fast-forward to half a billion years ago, and those plant cells containing chloroplasts developed into multicellular organisms and moved out of the sea. The first simple land plants still needed easy access to plentiful water for both reproduction and photosynthesis. But evolution and adaptation continued.

Today I can admire a beautiful flower with a woody stem clinging to a dry cliff more than a thousand feet above the freshwater sea.

The journey to this point was long and winding, but the view is truly amazing. Thinking about all this, I'm not sure that looking down was the best way to seek relief from strenuous awe.

Twinflower Faraway

Brooke Sapper, age 18

The sun rose slowly on its long midsummer trek across the sky as my friend and I sped west on Interstate 90 through South Dakota. Lush forests veiled by early morning humidity fell away, revealing cornfields with scattered, shady farmyards. At first the corn was thigh high; the ample rains and early spring had allowed it to race past the folk wisdom of "knee high by the Fourth of July." As if driving west was driving back in time, the corn shrank below the knee-high goal and ended up at ankle high before petering out into short-grass prairie with fewer and fewer scattered trees.

The humidity in the air also disappeared, and despite the vastness of our view over endless rolling hills, the horizon remained sharp. And on that horizon was a small mountain range with its tree cover so dark that from a distance the hills looked black. We'd arrived at the Black Hills.

Suddenly we were back among trees again, although dry ponderosa pine woodlands had mostly replaced the lush deciduous forests. Thickets of willows huddled along creek ravines, with birch, cottonwood, and bur oak providing a haven for birds. The Black Hills—rising like an oasis out of the prairie—are a biological mixing place, where species more commonly found in regions to the east, west, north, and south come together.

Swooping upward on a series of switchbacks, we arrived at Sylvan Lake in Custer State Park. Trading sandals for hiking boots and topping off our water supply, we struck out on Trail 9 to Black Elk Peak (formerly Harney Peak), the highest summit in the United States east of the Rocky Mountains.

The first section of trail was all a gradual climb. Afternoon sun glittered on shards of mica in the trail dust. Finally we began contouring around a ravine, and the trail became flat and shady. The respite from blazing sun and summer heat was welcome. And we weren't the only ones appreciating the Northwoods-like microclimate of this north facing cul-de-sac. A lush carpet of plants and flowers perched near eye level on the hillside.

I'd been stumped when trying to identify many of the dryland plants we'd seen elsewhere in the hills, but here I was among friends. Wild sarsaparilla spread its compound leaves over little starbursts of white flowers. Bunchberry was in full bloom with its pure white bracts (the showy part of the flower is not even its petals; its big bracts surround a cluster of tiny flowers) glowing in the shade. And rising from thick carpets of dollhouse-sized trailing vines, the pink, paired blossoms of twinflower gave the forest floor a Lilliputian look.

These little bell-like flowers creep under the radar of most hikers. They don't look like much from five feet up. But they did catch the attention of Carl Linnaeus, the "Father of Modern Taxonomy," who defined the scientific system of naming living things with a genus and species in 1753. Twinflower was Linnaeus's special favorite. Although Linnaeus was reported to be arrogant, and despite naming nearly 8,000 plants during his lifetime, the botanist refrained from naming any species after himself. Instead, he cheekily named beautiful plants after his supporters and christened weeds after his critics.

Linnaeus first named this sweet little flower *Rudbeckia* for two Lapland explorers who knew it well. Later on he also applied *Rudbeckia* to black-eyed Susans, which left twinflower in need of a new genus. Linnaeus's friend Jan Frederik Gronovius stepped in and named Linnaeus's favorite flower after the botanist himself. It became *Linnaea borealis*.

Linnaeus's influence circles the globe, so it's appropriate this his namesake plant does too. Twinflower grows throughout the northern hemisphere from Siberia to Sweden and across North America. In Europe, foresters consider twinflower to be an indicator of ancient woodlands. I often encounter it in the upper Great Lakes region, and I've also admired it in northern New England. The Rocky Mountains and Pacific Northwest also enjoy its presence. Despite it being so cosmopolitan, there's a definite gap in its range across the Great Plains. It can't survive those stark Dakota prairies.

So how did this tiny belle come to be separated east from west?

Bunchberry, thimbleberry, wild sarsaparilla, and more share this disjunct pattern in their populations. Not long after the glaciers melted, these plants would have enjoyed the cool, damp climate they prefer across much of the continent. Then about 9,000 to 5,000 years ago the climate warmed. During this Holocene Climate Optimum, warmer and dryer conditions forced these northern species out of the arid Great Plains. They persisted only in the refugia that mountains and boreal forests still provide. By vining perennially over its habitat, twinflower is able to survive in isolated microhabitats far from its strongholds. But climate change has become a significant concern for the conservation of this species, especially on the edges of its current range.

I felt a special kinship with twinflower on this hot, arid hike as we both found refuge in the cool shade of a north facing slope.

Wonderful Webs

Zachary Lellos, age 7

Beams from flashlights and smartphones bobbed around the Cable Community Center, poking into all the nooks and crannies, scanning window frames, and exploring under the eaves. Dusk had fallen quickly while we listened to Larry Weber—author of *Web Watching: A Guide to Webs and the Spiders That Make Them* (Stone Ridge Press, 2018), retired middle-school teacher, and ultraenthusiastic naturalist from northern Minnesota—narrate slides of local spiders. Now we were looking for the real things—not just photos.

Shouts of “Oh wow!” “What’s this?” “Good find!” and “Everyone should get a good look at this one!” cut through the evening air. Larry spotlighted a particularly messy corner. There, an unorganized jumble of threads glistened. “Cobweb spider,” he indicated, pointing to the tiny dot of a creature sitting in the mess. “You probably know some humans who keep their homes tidy,” he explained with a twinkle in his voice, “and some who don’t.” These spiders don’t.

You can often find cobweb weavers in dark corners inside and outside buildings. They have plump, round abdomens and usually hang upside down in their web—waiting for prey. During the day they may hide, so the best time to see them is after dusk. Earlier that afternoon when Larry led us on a spider hunt at the Museum’s Wayside Wanderings Natural Play Area, we had found several empty cobwebs in the bark crevices of a huge white pine.

It was a day packed full of spiders. The crown jewel of our afternoon spider hunt was a giant yellow crab spider, so named because of the way it holds its large front legs. In early summer Larry finds crab spiders on daisies, he told us, and they are white. By later in the summer, when the coreopsis, goldenrod, and other yellow flowers dominate, he finds mostly yellow crab spiders. They are the same spider. They just have the amazing ability to change color slightly to match the flower in which they hide.

Crab spiders can make silk (Larry demonstrated by dangling a smaller crab spider by its dragline), but they don't spin webs. Why bother when you can hide in a flower and wait for a pollinator to come to you?

However, a beautiful web can endear even the most dedicated arachnophobe to spiders for at least a moment. According to Larry the perfect web is on a two-dimensional plane, with not a single thread broken, and covered with dew. Those webs, with the radiating spokes, spiraling threads, and beautiful symmetry, are the creations of orb weaver spiders.

Each evening (or morning, depending on the species) the orb weaver will spin a new web, and a few hours later the web is gone again. What happens to it?

If the spider is successful, insects fly into the sticky, elastic threads of the spiral and get caught while struggling. The spider can rush forward and throw out sheets of swathing silk to subdue the insect. If the catch is large, the spider may wait patiently until the insect has exhausted itself in the struggle. Then the spider injects venom through its fangs to further subdue the insect, follows that with a dose of digestive juices, and finally slurps up an insect smoothie into its sucking stomach.

After all that, the web is usually pretty torn up. Indeed, many spiders will eat their web each day. The silk proteins are then recycled for use in their next web.

Larry's perfect web—with no threads broken—is an unsuccessful web with a hungry spider.

But you don't need perfection to enjoy spiderwebs covered with dew. Midsummer is when the spiderlings are hatching and growing, and many tiny webs may cover your lawn each morning. The funnel-shaped sheet webs you see in the grass are made by funnel weavers. If you watch closely, you may notice that the webs increase in size each week as the little spiderlings grow up.

Funnel webs aren't sticky, but they allow the spider, lying in wait at the base of the funnel-shaped retreat, to feel the vibrations of a passing insect, run out, and grab it. Funnel weavers aren't fans of cold weather, and they scurry indoors on cool nights. You may wake to find one trapped in the smooth porcelain jail of your bathtub or sink. They are harmless and easily moved back outside with a cup and piece of paper.

If you have bushes in your yard, you probably have "bowl and doily" webs made by the sheetweb weavers. These webs consist of a horizontal platform with many anchoring threads in a delicate maze above. The spider waits below the platform and may pull its catch right through the bottom of the web.

Late July is the perfect time to admire the delicate artistry of our eight-legged friends. Their wonderful webs catch mosquitoes, moths, and other pesky insects while adding breathtaking beauty to evening adventures and early morning walks.

Crossing the Moat

Mollie Kreb

One after another the group of adults on a field trip scrambled across the fallen tree that spanned the moat. Black, bottomless water licked at our boots, and armed sentries guarded the path with needle-sharp daggers. No, we weren't storming a castle; we were entering a poor fen.

Fens are a type of wetland with organic peat soils made up of poorly decomposed plants. While poor fens lack nutrients and are similar to bogs, there is one main difference: fens have some water flow (either from underground springs, a stream, or runoff), and that results in a neutral or basic pH. True bogs receive water only from rain, have no inflow or outflow, and are extremely acidic.

The glossy black moat we teetered across is a typical feature of these wetlands. Where runoff from the surrounding uplands contacts the bog or fen, it adds oxygen and nutrients and stimulates decomposition. Dead plants molder into slimy, insubstantial muck that gives way under floundering feet, and their leftover nutrients foster the growth of beautiful plants like water calla (related to the ornamental calla lily).

Once across the moat, we stepped gingerly onto the edge of a floating mat of vegetation and gauged its substantiality before committing our full weight. Although the moss sank and water welled around out boots, the carpet of interwoven plant remains held. One of the most fascinating things about bogs and fens is that they fill in from the top down. Shrubs and sedges anchor to the edge of an open pond and create scaffolding upon which sphagnum moss and more plants grow. In the anoxic, nutrient-poor water the plants don't decompose after they die. Instead, they sink lower as new plants grow on the substrate of their bodies.

Here, near the moat, the mat of poorly decomposed vegetation was only a few feet thick and undulated like a waterbed under our weight. It grew noticeably more stable as we walked toward the center of the wetland.

Sparky Stensaas, a naturalist, photographer, author, and our guide, stopped every few feet to point out patches of grass pink orchids, pitcher plants, and sundews. Glittering dragonflies and translucent damselflies zipped through the summer heat. “Keep your eye out for an elfin skimmer dragonfly,” he said hopefully. Jewellike cranberries ripened on their bed of moss. While the group paused to examine them, I got out my bog probe. This simple tool is just several four-foot lengths of chimney sweep extension rods. They are thin but strong and screw together securely.

The first rod was the hardest. I had to ram it forcefully into the thatch of vegetation before I felt the fibers part ways. After the second rod, resistance decreased as the lead rod found its way down through older and older peat. We held our breath as I screwed the fifth rod onto the set and pushed it down easily. The bottom was nowhere to be found. This told us two things: (1) Even near the edge of this fen the basin is more than 20 feet deep; (2) I need to order more extension rods. (Note: on a subsequent fieldtrip I brought along five additional rods, and we still did not find the bottom!)

Both amazed and disappointed, I began to extract my rods from the fen. As my hand grasped the second-lowest black wand, I gasped. It was ice cold! I know that the moss blanket on bogs and fens is an excellent insulator, and that bogs can harbor subsurface ice well into June, but on this 90-degree day, the surface of the bog felt like dishwater, and the contrast in temperature was startling. Those cold temperatures are one more cause of the slow decomposition here.

Without decomposition the resources that new plants need to grow are locked up in the bodies of old plants. To acquire nitrogen and other nutrients, several bog and fen plants have turned to carnivory. Pitcher plants and sundews are two of the most famous carnivorous plants. Their brilliantly specialized leaves capture small insects, and a combination of microorganisms and digestive juices break them down.

Beyond our probe location, near the edge of an open pool, we found great patches of small, sunny yellow flowers with a violet-like face and funny protruding spur. These horned bladderworts don't look dangerous, but theirs is the largest genus of carnivorous plants in the world. Under the innocent-looking flowers hide tiny traps.

Each trap, nestled among the threadlike leaves, is a bladder with a door that opens inward. The plant can pump water out of the bladder, flattening it and creating a vacuum inside. Bristles near the trapdoor look like a good feeding habitat and act to funnel prey toward their demise. When a minuscule invertebrate nudges trigger hairs near the door, the flap swings inward and sucks in both water and lunch. The door snaps shut as the bladder fills.

Digestive enzymes and resident bacteria digest the prey, which takes anywhere from fifteen minutes to two hours, depending on their size. To reset the trap, the plant uses special cells to pull nutrient-laden water into the stem. Not only does this feed the plant, it also restores the vacuum that is essential for the capture of future prey.

Recent studies show that a vibrant microbial community in and around the traps may also play a role in attracting prey and absorbing phosphorous.

Growing up, my imagination often wandered deep inside fairy-tale worlds. In this fen and others like it, I've found that those magical places truly exist. Dragons and damsels fly above a sparkling landscape where tricky traps lie in wait to catch unsuspecting travelers. The precious gems of flowers and fruits attract treasure seekers of all kinds. Cold danger subtends all that beauty. With this in mind, I can't think of a better way to enter and exit a fen than over a glossy black moat. Which is, of course, what we did as we hiked back to reality, swatting those dagger-wielding sentries the whole way.

Dreaming of an Elfin Skimmer

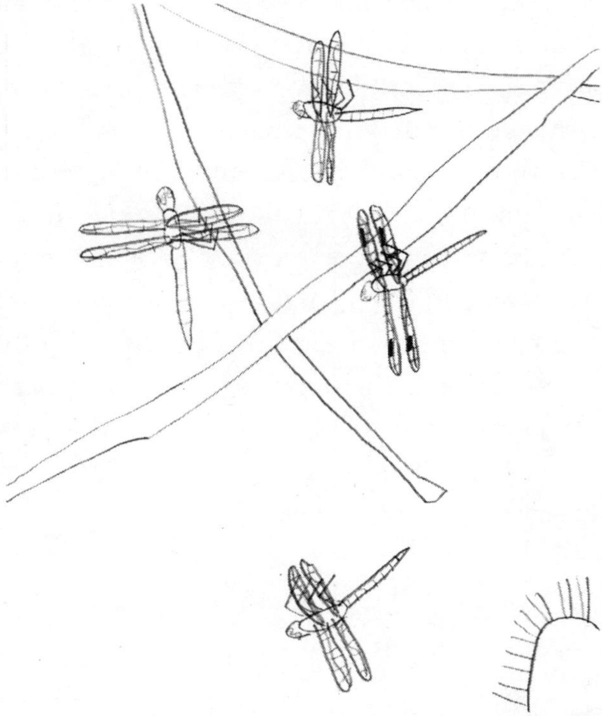

Wyatt Strapon, age 8

As we entered the soft mat of the fen, the starry white flowers of three-leaved false Solomon's-seal shone among puffy green clouds of sphagnum moss, and warm sunshine perfumed the breezes that wafted under my nose. I carried a camera. Kaylee wielded an insect net and backpack full of bug jars. As she began swishing the net through the tips of bog laurel and cranberry, I absent-mindedly mused, "I hope we find an elfin skimmer today."

At less than one inch long, elfin skimmers are the

smallest dragonflies in North America. That superlative, plus the fact that they inhabit the magical spaces of bogs and fens and have a delightful name, has endeared them to me. But I've only known of their existence since last July when naturalist and author Sparky Stensaas led a field trip to a nearby fen and also expressed hopes of seeing one. To be honest, although they've zipped around in my imagination since that day, I've never seen one, and I couldn't even tell you how to identify them in the field. They'd be shimmering and fairylike, I presumed, like my wistful but not exactly rational request to the universe that we find one.

My official, stated purpose for the day was to offer assistance with plant identification and logistics at a BioBlitz. A BioBlitz is an intense period of biological surveying in an attempt to record all the living species within a designated area. For most BioBlitzes the public is invited to join in the fun to learn more about biodiversity (including dragonflies with fanciful names) and better understand how to protect it.

Kaylee and I had rounded up 13 experts and participants who were interested in exploring Sugar Bush Fen in Northeast Lake State Natural Area. We'd all doused ourselves in bug spray before taking off through the woods and onto the soft sphagnum moss mat of the fen.

Into Kaylee's bug jars went flies, dragonflies, butterflies, beetles, bees, grasshoppers, mosquitoes, and a single spider. Into my camera went photos of pitcher plant flowers (their petals still dangling like burgundy silk!), puffy tufts of cottongrass, the long-petaled beaks of cranberry (crane-berry), and the dainty pink shoes of lady's slipper orchids.

The fen was a dream. We caught a tiny spring peeper in a bug jar and admired the frog's suction-cup toes before putting him back on the hummock from whence he came. We slogged through the shadeless, boot-grabbing sedge mat to reach the enchanted pool of liquid darkness at the center. Water lilies dotted its margins, and tiny thickets of sundew too. Each spoon-shaped leaf bristled with tiny pink hairs, and every luminous hair sparkled with a drop of either sticky, sweet nectar or deadly digestive enzymes at its tip. A few leaves were still digesting a six-legged lunch.

The graceful, nodding stems of fen sedge brushed our muck boots as we finally exited the wonderland ready for some shade and a shower.

Back at the Cable Natural History Museum, Kaylee's real work as curator began. She started by emptying out the jumbled, bug-filled jars, spreading the wings of dragonflies and butterflies so that they could dry properly, and pinning the smaller insects on white foam. With the specimens preserved, she could then turn to the task of identification. For my part, I uploaded plant photos to the iNaturalist website and tagged them as part of the CNHM BioBlitz 2017 project, which anyone can explore.

I was at my desk deep into computer work when Kaylee's voice preceded her around the corner: "We found an elfin skimmer!"

His tiny body was pinned in one corner of her white

foam square. Having finally done my research, I could tell by his pale blue, pruinose (frosted like a grape) abdomen that this male had matured fully after a shiny black 10-day adolescence. In life he would have patrolled a small territory surrounding a single puddle in the fen's floating mat. Kaylee probably netted him from the bristle of sedges near the fen's open pond.

Not being a strong flyer, he would have perched—clear wings drooping forward—when not chasing off rivals. After mating with a female, he would have guarded her as she deposited eggs into his pool with dainty taps of her brightly striped abdomen. Female elfin skimmers look quite different, and their black and yellow markings are thought to imitate wasps and provide some protection from predators. Plants don't care about warning colors, though, and the skimmers' size makes them vulnerable to becoming mortally entangled in a sundew's trap.

Now this elfin skimmer is perched on a pin for perpetuity. His mortal trap was Kaylee's net. He's no longer able to take shimmering flights through my dreams. Looking through the microscope at his crystalline wings, frosted body, and sparkling brown eyes, I feel like I've been given a precious gift. And yet, by pinning down a glittering zip of imagination and making it real, I feel as though I've lost something too.

Pitcher Plant Partnerships

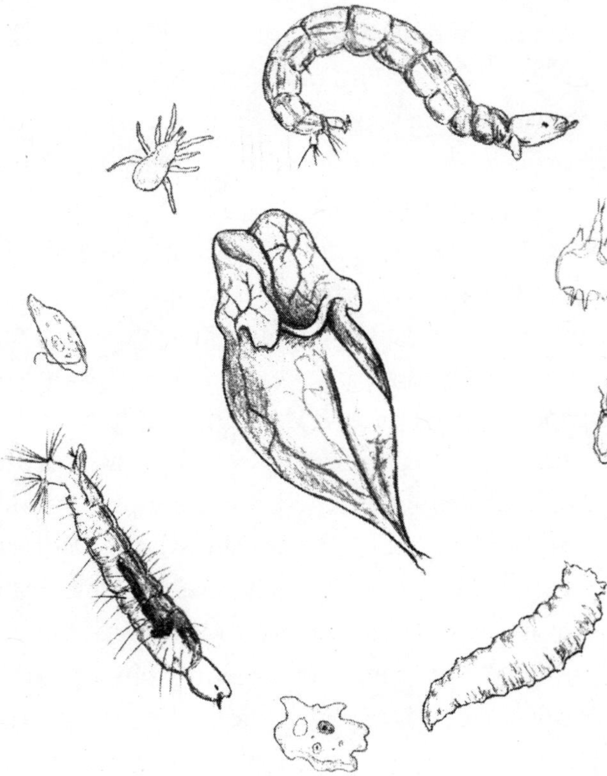

Mollie Kreb

The glossy black water in the moat around the fen licked my toes again this week. Today, after I stepped onto the relative security of the spongy bog mat, I reached back to heave a camera bag across too. Teetering along the branch bridge came two cameramen, a sound guy, a geologist, a few bystanders, and three teenage TV stars from the nationally syndicated *Aqua Kids*.

During their Great Lakes adventure, these kids wanted to see a bog. We decided that a poor fen was close enough, and I knew of a lovely one to show them. Soon after we'd crossed onto the firmer mat of vegetation beyond the moat, someone almost stepped on a beautiful clump of oddly shaped leaves. One of the kids exclaimed in wonder and asked me if this was indeed the famous pitcher plant. It was, so of course we now had to get the same question on camera. Backing up a few paces, we staged a meeting in the fen by walking in from opposite directions and feigning surprise. Our paths crossed perfectly at the elegant clump of pitcher plants with three tall flowers and a bouquet of red-veined leaves.

As I explained that the acidic, low nutrient habitat of bogs and fens drives the pitcher plant's need for nitrogen and its resulting carnivory, I poured the contents of one water-filled leaf into a clear vial. I may have squealed in excitement. Two big flies floated at the top—one dead and the other mostly dead. Those were soon to be prey; no big deal. Near them hovered a half-inch-long, pale, segmented sea monster. It was this larva that triggered my glee.

I love the experience of reading about something new and then finally encountering it for real. It almost feels like a dream come true. I'd recently learned about this creepy tenant of pitcher plants in the wonderful book *North Shore* by Chel Anderson and Adelheid Fischer.[2] Now here was the critter right on cue.

Flesh fly larvae hatch while still inside their mother, and

she deposits one wiggling baby in each pitcher plant leaf. If by accident two larvae end up in the same tiny pool, well, let's just say that one leaf ain't big enough for both of them.

Flesh fly larvae act as pitcher plants' teeth. Like Cookie Monster, they shred their food ferociously and let the crumbs fall where they may. Luckily, a cleaning crew is on the job. Tiny mosquito larvae of a species that only lives in pitcher plants filter-feed on crumbs suspended in the pool, surfacing occasionally to breathe. The habitat in a pitcher plant pool is so rich and rewarding that larvae store up extra energy. The females of these well-fed mosquito larvae metamorphose into adults who do not need a blood meal to lay eggs. Now those are mosquitoes I can appreciate.

We all oohed and aahed at the contents of the vial for a moment, and then I pulled out my knife. Carefully I slit the pitcher plant leaf all the way from its slippery, nectar-laden lip, down the red-veined, slightly sweet side, to the narrow tube at its base. Ewww! We shouted in delight as a black mass of dead insect parts writhed in the sun. The dead stuff wasn't moving, of course, but their decomposers were.

Down here where the leftovers settle, midge larvae draw their oxygen directly from the water. With powerful mandibles they break down the insect bodies even further, releasing more crumbs to the mosquitoes and plenty of nutrients to the pitcher plant.

This trifecta of larvae is a classic example of resource

partitioning and finding one's niche, but there's even more to the story. A whole community of microscopic rotifers, mites, protozoans, and bacteria feed on the leftovers and on each other. You might think the pitcher plant would resent all these freeloaders. Instead, a high diversity of organisms increases both the digestion rate and the availability of nutrients. The pitcher plant absorbs plenty of good stuff through special cells deep in the bottom of its leaf cup.

The kids, the cameramen, and I soon moved farther into the fen, exploring the dainty yellow flowers of horned bladderworts, the glistening leaves of sundew, and tart bog cranberries. Throughout the day I was impressed by their genuine enthusiasm, intelligence, and humor.

Exhausted, I finally retreated to the shade to share some parting words with the camera. "Thank you, Aqua Kids," I smiled, "for helping me get even more people excited about bogs and fens!" Just like the critters in a pitcher plant, we made a great team.

The Brookies of Cap Creek

Ayla Baussan, age 8

"Stick a toe in the water," dared Wisconsin Department of Natural Resources (DNR) fisheries biologist Max Wolter with a sly grin. The teenage boys (and one mom) who had joined me on this adventure hesitated for only a second before taking the dare. "Wow, that's cold" was the general response. We'd just dipped our feet into Cap Creek, a spring-fed tributary of the upper Namekagon River, and that cold water was part of what brought us here. The cold water also attracts native brook trout, and this little stream hosts one of the best brook trout habitats on the Namekagon.

Each year the DNR collects data on the number and size of trout in this stream. The researchers accomplish this not through the elegant casting of a fly rod, like we attempted a few weeks ago, but with wands that send currents of electricity zipping through the water. Temporarily stunned fish can be scooped up from the bottom with nets (fish don't float when stunned, only when dead and decomposing). Fish shocking is a common research technique—and one that holds quite a bit of intrigue for kids of all ages.

When we'd arrived at the Cap Creek Landing on the Namekagon River, the three DNR biologists were sitting there with their little research boat, counting and measuring fish. Max looked up apologetically, saying, "This might take awhile; we caught an unusually high number of trout in our first sweep." So we stood there watching and listening as two biologists pulled fingerling trout out of their nets, dropped them one at a time into measuring troughs, and called out lengths to Max, who recorded the numbers on his data sheet. While a bit tedious, this is the reality of science in action.

Things got more exciting when Max brought out a small aquarium on a tall pole. He sank the pole in the sand and then grabbed good-sized brook trout out of a net. Now at eye level, we could admire the beauty of this incredible fish—the only truly native trout in the Namekagon. A pattern of yellow and olive-green markings was overlain by scattered red spots, each with a halo of pale blue. The pectoral fins were a lovely shade of orange. I've never seen a more beautiful fish.

According to Max, they usually only find a dozen or so

trout in this section of the Namekagon River, right below the mouth of Cap Creek. Today they found close to a hundred. Of course we asked why. "The mild winter," he replied immediately. "All the little ones survive the winter better when it's short and there's not as much ice buildup." He continued, "It's neat to have long-term data, because you can look at how the weather impacts trout. In all the really long winters we had bad trout numbers. In all the short winters we had good numbers. You can start to see patterns."

Of course, it's not just the length of winters that affects trout populations on the Namekagon and its tributaries. "Trout are fragile," remarked Max. "They need cold, pure water, good habitat, and few contaminants." In the old days the Namekagon River provided all that. Way back in 1831 the explorer Henry Schoolcraft noted the abundance of brook trout in this river, even in the warm summer months. Later, in 1883, as tourism became a local industry, one fly fisherman claimed that "all trout streams in the state must yield the banner to the noble and lordly Namekagon."[3]

The logging boom of the late 1800s and early 1900s changed that. With the large trees felled off the Namekagon's banks, sunlight warmed the river. Extra sediment washed into the channel and covered some of the cold water springs that used to moderate the river's temperature. Woody debris that previously provided cover was removed from the channel. Logging dams pooled water and allowed it to heat up before flowing downstream. None of those changes are good for a fish who can't survive in water above 68 degrees.

Cap Creek had its own set of challenges. In the 1950s a trout farm diverted the water of this spring-fed stream into several excavated ponds. Although trout thrived in the hatchery, no native fish frequented the creek. In 1988 the National Park Service acquired the trout hatchery and began major restoration work. Cap Creek was restored to its original channel. Extra sediment was removed, revealing groundwater springs that today we watched bubbling up from the sand. They help keep the stream at a steady 50 degrees all year long. Logs were placed in the channel to provide cover and shade for the fish. "They work," said Max, pointing to one of the logs along the bank. "We got seven or eight adult trout out of that spot when we came through with the nets." He continued, "This is a great example of what you can do to restore something."

Native prairie plants higher on the bank swayed in the breeze. They were part of the restoration too. But all along the water's edge were carpets of forget-me-nots, a European species that has escaped and spread along many stream banks. I commented on it being an invasive. "Well," spoke up one of the biologists, "invasive or not, it's good cover. That's where all little fish were hiding when we came through." "Plus it's pretty," noted one of the boys.

This stream has another introduced species who may or may not be a problem. European brown trout were stocked here starting in 1883. They do better in warmer water, and therefore frequent the main river channel more than the chilly waters of Cap Creek. The researchers only caught a few brown trout on their sweep up the creek.

By some accounts brown trout eat the young brookies and push them out. But with the changes in water temperature brought about by logging and the uncertain future of climate change (which predicts temperatures in northwestern Wisconsin to be more like eastern Kansas by the year 2095), brown trout may have a role to play in the ecosystem. The interagency Fisheries Management Plan for the Namekagon and St. Croix Rivers describes the introduced brown trout as "an ecological surrogate for brook trout" and states that "brown trout are now a keystone species, maintaining the basic biological integrity of this fish community."

Sticking our toes in the chilly stream today, we touched a bit of a common current that runs through the history of North America. It flows from abundance to exploitation, habitat loss to restoration, and now preparation for an uncertain future. Conserving and creating a world that we—and the trout—will like living in is an upstream battle.

Forest Bathing

Diana Randolph

Through the soles of our feet we experienced the transition from gravel drive to leafy duff. Conversation faded as we entered the forest on a faint path and then formed a circle in a natural foyer surrounded by slender saplings. Fourteen women stood quietly while I read a poem. Above us, maple leaves glowed against a gray sky still lingering after a midnight rain. As the last line concluded, a gust sent a crescendo through the canopy.

We closed our eyes and inhaled deeply, feeling the goodness of air as it filled our lungs. As cool sensation rushed in, we imagined all the previous lives that the carbon, oxygen, and nitrogen molecules had made possible. Returning our breath to the world, we sent our carbon back out into the forest to fuel new life. Giving. Receiving. Giving. Receiving. The breeze swirled all our exhalations together and carried them off through the trees.

More than simple air flowed into our lungs, though. All around us, nature was giving off an array of chemicals. These phytoncides include a bouquet of volatile organic compounds released by plants. Their main purpose—from the plant's perspective—is to prevent it from being eaten, infected, or decomposed. Appropriately, the word *phytoncide* means "exterminated by the plant." These toxins are categorized as secondary metabolites, or chemicals that aren't essential for normal growth and reproduction but often help the plant survive in other ways. The strong smells and health benefits of onion and garlic are derived at least in part from phytoncides. While all plants have some, tea tree, oak, cedar, locust, and pines are known for having high levels of these helpful compounds.

For the plant to prevent damage to its tissues from too high a concentration of its own toxins, it has to have a way to excrete the excess and maintain balance. Some may escape to the air when pores open to let in carbon dioxide. Others may leach out and flow away with a shower of rain. Still more are released as the forest duff decomposes. However it happens, these phytoncides become part of the forest at large, and we breathe them in.

It's a wonderful gift. Not only do the antibacterial and antifungal properties of phytoncides help plants fight disease, they also stimulate our human bodies to increase the number and activity of cancer- and virus-fighting white blood cells (also called natural killer cells) and to decrease the concentration of stress hormones.

That breathing deeply in a forest is good for us comes as no surprise to most people. Anyone lucky enough to live near the woods and smart enough to take time to enjoy them feels the benefits.

When I tell people that it has a name, though, they usually think I'm kidding. The Japanese Ministry of Agriculture, Forestry and Fisheries wasn't joking, however, when in 1982 they coined the term *shinrin-yoku*, which translates roughly to "forest bathing." In a country where they have a special word for "death by overwork," this stress-reducing preventive medicine has become extremely popular. On official forest bathing trails, visitors regularly submit to measurements of blood pressure and stress hormones as researchers gather data. Psychological research reveals that forest bathing reduces anxiety, depression, anger, confusion, and fatigue while boosting cognition, focus, and empathy.

It's not just the phytoncides. According to scientists, the air in forests and near water has relatively high levels of negative ions, which are purported to boost our mental outlook. Also, simply looking at trees is good for us. And then there are the bacteria.

Having healthy gut bacteria is emerging as a significant

factor in mental well-being. In another example, inhaling the common soil bacterium *Mycobacterium vaccae* has been shown to increase serotonin levels in the brains of mice. Not only does this "happy chemical" decrease anxiety, it also makes the mice smarter! Mice given the bacteria navigated a maze twice as fast as the control mice. The effects do not last long, though, and scientists surmise that humans would need to be exposed about once a week to reap the benefits of these healthy bacteria.[4]

That's fine with me. It was lovely to walk among the regal hemlocks of Fairyland State Natural Area that morning as patches of blue sky peeked in from above. We offered words of gratitude. I saw the lines on the foreheads of the women soften. Peaceful smiles glowed. Inhale. Exhale. Receiving. Giving. Once again, the breeze swirled all our exhalations together and carried them off through the trees.

Northwoods Lobsters

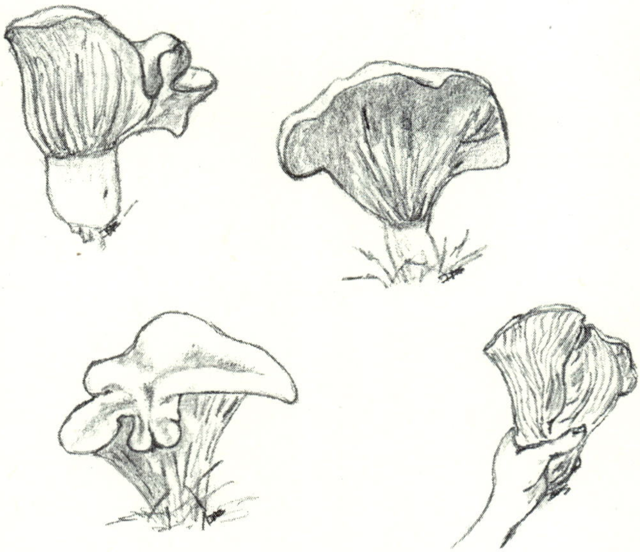

Doug Stubbe

After many days without rain, the dry spell was getting monotonous. So when the gentle hiss of liquid on leaves permeated my kitchen window screens, I donned a bright yellow slicker and headed out into it. A slow, quiet walk in the damp woods has a poetry all its own.

Colors were muted in the gray dusk, but scents were made vibrant by the splash of drops sending dust molecules skyward, where the damp air stuck them to the insides of my nose. I breathed deeply, again and again, on my way down the driveway. Sweet, green raspberry leaves. The tang of the bog. Wet asphalt. Wet gravel. And then, the earthy bouquet of the woods. This was forest bathing at its best.

While focusing on my nose, I let my eyes wander in that fuzzy, nonseeing way.

Until, that is, a flash of bright orange switched me between senses.

The oddly shaped blob of color right next to the trail didn't look like much. But I knew instantly that it was treasure. In fact, it was a lobster—right here in the center of the continent! Not a crustacean, mind you, but a mushroom.

Lobster mushrooms, like many good things, are the result of a relationship between two living beings. The orange-colored fungus, *Hypomyces lactifluorum*, parasitizes the mushroom of another fungus, and in doing so creates a tasty treat. Neither of the common host mushrooms is edible on its own. While not poisonous, *Lactarius piperatus* is reported to have an intensely peppery flavor. *Russula brevipes*, the other common host, is also harmless but has a crumbly, Styrofoam texture that would be unpleasant to eat. *Hypomyces* not only renders a pleasantly firm, dense texture, it also neutralizes the peppery flavor and imparts an interesting seafood flavor.

Lobster mushrooms are a delight for beginning mycophiles, since they and their hosts are easy to identify. They are on the short list of mushrooms that I'll eat without expert help. The outer surface is bright orange and looks slightly pimply—those dots are the reproductive structures for the parasite. Breaking open the gnarled mass of what used to be a mushroom cap, I found an appealing white center.

Continuing down the trail, I started to notice plain white mushrooms all over the woods. "*Puhpowee* was here," I thought to myself. An Anishinaabe word, *puhpowee* means "the force that causes mushrooms to push up from the earth overnight." These must be the unparasitized cousins of my lobster. But which host were they?

The first crumbling snap of the stem made me think of a Styrofoam *Russula*, but then I noticed milky sap oozing from a broken gill under the cap. This *Lactarius* was lactating. "So how peppery is the milk?" I wondered. After debating the risk of eating a raw wild mushroom (something that is not advised), curiosity won out and I touched the tip of my tongue to the milk. It was spicy, but not any worse than a peppercorn stuck in your teeth or an extra dose of wasabi on your sushi. Eating a full bite, though, would have been painful.

With tongue burning, nose humming, eyes flashing, cool raindrops tickling my bare knees, and the patter of rain filling my ears, I hurried back down the trail toward my kitchen. Although I've known about the edibility of lobster mushrooms for several years, I had yet to taste one.

Soon my favorite cast-iron skillet was swirling with butter, and a pile of fresh garlic was mounded on the cutting board. Into the pan went the bowl of cut and cleaned lobsters. Into my nose rose a savory perfume. Into my mouth went the first hot morsel with its bright orange rind. Without garlic, it tasted bland but pleasant. It was a nice experiment but not dinner. In went the pile off the cutting board. The flavors of garlic and butter warmed my cool evening as the rain drummed harder on the leaves.

Lobsters in the Northwoods. After rain, after many days without rain, you never know what you might find.

Growing Up Before My Eyes

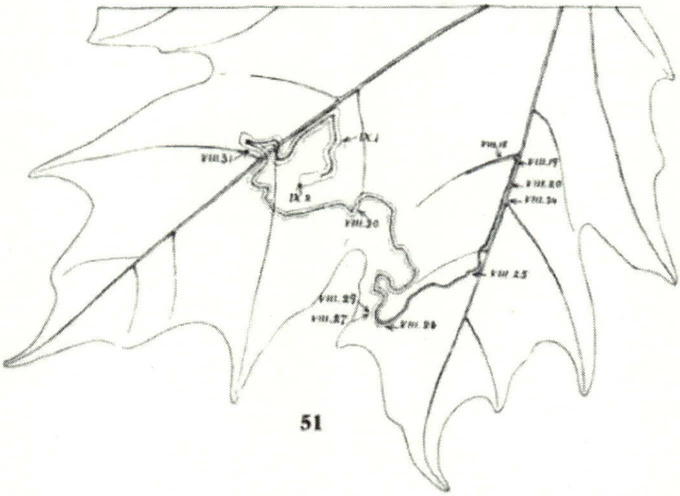

51

Annete F Braun

Water still dripped from the trees onto saturated duff, but blue sky peeked shyly through the clouds. Tired of the rain, my parents and I had taken advantage of a short break to go for a walk. Partway up the gravel road I followed a couple of mushrooms into the woods and then looked up. Above my head dangled a yellow and green sugar maple leaf with a squiggly line wandering through it.

Immediately enamored (who wouldn't be?), I plucked it down for a closer look. You may not think that animal tracking extends to insects, but it uses the same process of looking for clues that eventually tell a story. The squiggle on this leaf was typical of trails left by leaf miners, a group of insects whose larvae feed on the insides of a leaf between its "skins." Some internet research and a quick email to Charley Eiseman, author of the fantastic resource *Tracks and Signs of Insects and Other Invertebrates* and the BugTracks blog,[5] turned up a narrative we can all relate to.

The story begins near the base of the leaf where the trail of the leaf miner starts out narrow, about the width of a line from a felt-tipped pen. The line itself is composed of a dark center stripe and paler, translucent edges. It records the growth—essentially the childhood—of a moth larva of the species *Glaucolepis (Trifurcula) saccharella*.

The egg was laid earlier this summer, maybe about a month ago, by a tiny brown-and-white moth—wingspan four millimeters. The larva hatched, burrowed its miniscule body between the upper and lower epidermis of the leaf, and began to eat. What is there to eat between the skins of a leaf? A juicy substance called mesophyll fills this inner space. It is full of chloroplasts, the green engines of photosynthesis.

Like a mountain path, the narrow trail of the leaf miner scribed switchbacks in the lowest lobe of the leaf near the petiole before widening and flowing like a river toward the leaf's tip. It was fascinating to follow that widening trail and imagine the larva growing bigger as it fed.

The line of frass (caterpillar poop) down the center of the trail tells its own story. Through much of the mine the frass is just a fuzzy brown line, but in some sections a zigzag pattern becomes visible. It looks for all the world like one of the fancy stitches on my sewing machine. Charley Eiseman explained the cause of this pattern in an email. “It’s mesmerizing,” he wrote, “to watch a nepticulid larva at work under magnification—the central frass line is made up of many tiny pellets, and the larva is constantly depositing them as it eats, wiggling its tail end around methodically to place them just so.”

When I read that, you can bet I added “watch a leaf miner larva at work” to my bucket list!

Looking at my leaf, I can imagine the absolute concentration of the larva as it went about its essential business of eating and necessary business of pooping. Occasionally, though, the pattern seems to break. There’s a slight accumulation of frass, then a thinner line, and—is it my imagination?—the entire mine seems to get a tad wider. I suspect that these interruptions in the pattern indicate places where the caterpillar grew too big for its skin and molted, becoming a new instar.

Just before the end of the mine, its color changes drastically. I noticed it first on the LCD screen of my camera. I had taken a photo of the leaf backlit by the sun. Most of the mine lit up beautifully, like a stained glass window. Near the tip of the leaf, though, the mine around the frass line became invisible! Turning the leaf over, I noticed that the color on the back side also changed drastically at that point. For most of the mine, I could see that a layer of leaf was more intact on the lower side of the mine than the upper. Then for the last stretch of the mine, the insides of the leaf seemed to be completely hollowed out. The frass line appeared thinner and straighter too, like the path of someone on a mission. My hunch is that on the final day before its metamorphosis, the larva suddenly became a ravenous teenager and began eating everything in reach, and eating it so fast that the frass line stretched thinner, like a line from a pen swooped lightly across a page.

Finally, maybe a quarter inch from the end of the mine, the frass line ends. The mine continues as a transparent section of leaf missing all its green. This, Charley confirmed for me, represents the length of the larva—frass marking one end, uneaten leaf marking the other end—just before it popped out of the leaf, dropped to the ground, and formed a cocoon to overwinter before completing its metamorphosis to a moth next spring.

"Watching" the process of this larva growing bigger and bigger and finally leaving home to metamorphose reminds me of all the little old ladies who would pinch my cheeks lovingly and admire how I'd grown. Then they'd hold up their thumb and first finger about an inch apart and exclaim, "I've known you since you were this big!" If I could find that tiny larva, now wrapped tightly in its cocoon, you can bet I'd pinch its cheeks too! After all, I feel like I watched it grow up before my eyes.

The Channels of Bark Bay Slough

Alyssa Core

We could not have asked for a more perfect morning. The black tannin-stained waters of the Bark Bay Slough were perfectly calm under a warming sun as we launched my trusty Old Town Penobscot 16 canoe from the dock. We paddled easily, blissfully, into the morning, savoring the magic of floating through a beautiful place.

Bark Bay nestles between two ridges of erosion-resistant sandstone on the western face of the Bayfield Peninsula near Herbster. It is the estuary of the Bark River, and the back of the bay is almost entirely cut off from Lake Superior by a giant sandbar. Boggy little islands dot the enclosed slough—each with their own artistic arrangement of weather-sculpted bonsai trees. Exhibiting both scrappiness and grace, the stunted white pines, black spruces, and tamaracks twist toward the sky. Around their bases crowd thickets of leatherleaf and sweetgale dotted with wild roses.

As my cousin Alyssa and I paddled slowly eastward through the island maze, mats of bog and fen vegetation crept out from the shore and narrowed our passageways. We never knew if our next turn would bring us to a cul-de-sac or a channel. That element of mystery was delightfully thrilling and brought back memories of playing pretend with Alyssa in my grandpa's shrubbery. Add to that a shy painted turtle poking his head out from under a water lily leaf, and we were completely entertained. I don't often have a day where my curiosities so perfectly match that of my adventure buddy. On this day, I savored every breath.

Following our interest, and beckoned by their beauty, we spent at least 10 minutes in a patch of watershield, *Brasenia schreberi*. The palm-sized oval leaves of this aquatic plant float on the surface of calm waters. A thin, flexible stem attaches to the middle of each leaf's underside and tethers it to a horizontal stem that runs along the muddy bottom. Clear, slippery jelly coats the undersides of the leaves and protects them from grazing snails and harmful microbes. When it grows in thick patches—which is often—watershield creates a spectacular mosaic of color, especially in late summer when brilliant bursts of red and yellow paint the dying leaves.

Other patterns add interest to a patch of watershield too. Most of the glossy leaves are riddled with holes or dissected by squiggly lines, which all look black against the tannin-stained water.

What would make those marks? And was it still around? One by one, I held a few different leaves up to the sun and tried to capture this living example of stained-glass beauty. To my surprise, these trails differed from other leaf mines I've seen. Leaf miners are a group of insects (including moths, sawflies, flies, and beetles) whose larvae feed on the insides of a leaf between its "skins." In some cases the tracks on the watershield leaves had been worn clear through, with only a few of the tougher veins keeping the sides of the seam from ripping apart.

Another more subtle difference was that although not all the trails were the same thickness, each trail was the same width from beginning to end. What I've enjoyed about most other leaf mines is that they start small and get bigger, allowing me to imagine the little larvae growing up.

Perhaps the only person who's investigated the origin of watershield's leaf marks was a fisheries biologist named Adelbert L. Leathers in 1922.[6] From his writing I learned that the squiggles are actually channels and not mines. In a mine, the larva burrows between the upper and lower skins of the leaf, eating the innards but leaving both of the outer layers intact. In contrast, these channels in the watershield were created by removing the upper epidermis entirely.

With weak mouthparts, the midge larva *Polypedilum braseniae*, cuts off a strip of the upper epidermis, scrapes it clean of nutritious green cells, and then uses silk to fasten the strip into a roof above its channel. This setup allows water to continuously flood the channel. The larva breathes through blood gills and will die if it dries out. Even submerging too deeply will kill the larvae, presumably because they need the superoxygenated water found near a pond's surface and in contact with living plant cells.

Mr. Leathers watched as green plant cells moved through the digestive tract of the pale yellow larvae. He noticed that when the food exits the other end as frass (poop), it's still useful. The larva places frass where it can help hold a strip of epidermis to make it easier to scrape, and it eventually becomes part of the channel roof. After seven to ten days, the larva spins a silken pupal case and metamorphoses into a tiny midge with shimmering wings and an ethereal green abdomen.

Much mystery remains. Although Mr. Leathers was able to find the midge's eggs on submerged water lily leaves and successfully hatch them, he could not get the tiny larvae to grow or survive. They arrive at the watershield leaves to begin channeling only after they are midway through their "larvalhood." The strength it takes to cut through the epidermis requires a certain minimum size of larva, and the larvae would rather start in an old channel and branch off than go to the work of sawing through fresh material.

I love discovering that there are unsolved mysteries in science. When I start investigating something new to me,

I never know if my next question will open up to a cul-de-sac or a clear channel. That element of mystery is delightfully thrilling.

Tongues of the Earth

Donna Post

Recent rains watered the Earth like one of those old terracotta chia pets, and now mushrooms sprout from every sodden surface.

In the breathless shade of a hemlock forest, I wandered among the buzzing mosquitoes and stoic trunks, fairly tripping over fungi. The trees dripped with a thunderstorm's leftovers, and (looking forward to snowflakes instead of bugs) I stopped to catch a drop on my tongue.

Bright orange globs of hunchbacked lobster fungus, sunshine-yellow spikes of club fungus, and the glowing white cap of a destroying angel illuminated dim, shadowy duff. Without a guide—human or paperback—I simply admired the diversity of colors and forms, identifying only a few of the most distinctive organisms from my tenuous memory.

In these forest soils beneath my boots, 90 percent of the living biomass is fungal. An interwoven mat of mushroom mycelia connects trees with resources, returns nutrients to the soil, and exchanges information. Some call it the "wood wide web." Mycologist Paul Stamets calls it "Nature's Internet." I see a fascinating parallel with my own internal microbiome: For every human cell in my body, there are about 10 resident microbes. They help me digest food, tune my immune system, and even alter my mood. None of us is ever truly alone.

Mushrooms make pretty good company, I would say, and perhaps that is because mushrooms are more closely related to animals than they are to plants. For one, plant cell walls are made of cellulose, while mushroom cell walls are made of chitin. Chitin is the stuff of crab shells and butterfly wing scales. It doesn't exist in plants. Plants are autotrophs—they make their own food out of sunlight, water, and air. Fungi and animals are heterotrophs who get their energy by breaking down food they find. Plants exhale oxygen; fungi and animals inhale it.

Despite the incredible community of life here in the woods, this magical place was a source of solace and solitude for a young Mary Griggs Burke. Out of her love of Fairyland (its official name) grew an admirable land ethic. She wrote, "I am glad more people are enjoying the beauty of nature—but only if people value the land and treat it with respect can they preserve what they enjoy."

And with that in mind, she founded the Cable Natural History Museum.

These woods are full of connections. I wouldn't be here without Mrs. Burke's foresight, generosity, and love of nature. The trees would not be here without the mycorrhizal fungi that extend their root systems. Each tree has hundreds of thousands of miles of fungal threads associated with its roots. The fungi supply the tree with nutrients and water. In exchange, the tree provides sugars synthesized with help from the sun.

Scientists are discovering that sharing—symbiotic relationships—among highly diverse and unrelated organisms is more important than they used to believe. "Life is a network of cross-kingdom alliances," says theoretical biologist Lynn Margulis.[7]

A few days later, when I brought a multigenerational family into Fairyland with me, it only seemed to multiply the magic. More eyes saw more fungi, and more people exclaimed exuberantly over mushrooms too. While Rachel Carson wrote that "if a child is to keep alive his inborn sense of wonder, he needs the companionship of at least one adult who can share it, rediscovering with him the joy, excitement, and mystery of the world we live in,"[8] I've found that adults need such a companion as much or even more.

So, we discovered things together. This time I carried the *Lone Pine Field Guide to Mushrooms of Northeast North America*. From there we learned that the tiny red-orange cups sprouting from a rotting log are named eyelash fungus. When we leaned in closer, the fringe of black "eyelashes" on the margins became visible. Chuckles of appreciation rippled through the group.

The next find was a patch of little clubs with fleshy, creamy white fans atop darker, somewhat velvety stalks. The name velvety-stalked fairy fan seemed amusingly appropriate. We don't need to know the names of everything to appreciate them, but sometimes it does add to the fun.

"Look here," whispered the family patriarch with a wiggle of his toe. At the end of his shoe we found an elfin grove of two-inch-high black clubs. They blended in so well with the shadows and damp wood that we were quite proud of ourselves for spying them. But the name common earth tongue seemed like an insult to our keen observation skills. Nearby, a cluster of similar but yellow clubs sprouted cheerily in the gloom. The name yellow earth tongue endeared us a bit more.

Tongues of the earth indeed. These fungi do more than just look like they might lick the ground. Animal-like in many respects, they digest almost every bit of the earth at one time or another and make the rest of our lives possible in the process.

Mysterious, and mostly hidden, the fungi perpetually beneath our feet are often forgotten. In these rainy days when we feel like our skin might either mold or get sucked dry by mosquitoes, the mushrooms thrive. Perhaps the Earth is, in essence, sticking out its tongue—its earth tongues—to catch the raindrops.

Bear Corn

Drew Guttormson

"Stopping!" I shouted to the riders behind me. We all skidded to a halt where the singletrack mountain bike trail crossed an ATV trail. I'd spotted something weird in the woods and wanted to check it out.

The Esker Trail, near Cable, Wisconsin, provides some pretty amazing scenery and a great adventure whether on a mountain bike, a hike, snowshoes, or a fat bike. About 14,000 years ago, near the end of the last ice age, this trail was buried under a mile or more of glacial ice. Rivers of meltwater coursed along the glacier's surface and plunged downward through crevasses or through vertical shafts called moulins.

The rivers of meltwater didn't disappear, though. After one poured down a moulin, it began to flow under the glacier. Punky ice above was easier to erode than frozen ground below, so the river created a tunnel for itself in the glacier. Snaking along, often with several sinuous channels braided together, this subglacial river dumped tons of sediment into its riverbed. In response, the river's tunnel stretched higher, and the water deposited even more sediment. Once the ice was gone—having gushed away in a torrent of melt—it left a long, sinuous, steep-sided ridge, which we call an esker.

Which happens to be a great place for mountain biking.

Eskers make great habitat too. Pines and oaks thrive in the well-drained soils. We biked through grove after grove of these trees, with glossy, ripe blackberries lining the trail. All those oaks occasionally attract a weirdo, though, and that's what caught my attention. Dark, scabby blotches marred the pale-khaki leaf litter. Based on a series of nature identification requests I'd had recently, I had a good guess about their origin.

Bear corn, also known as American cancer root, or *Conopholis americana*, is a strange-looking plant. For one, it doesn't have chlorophyll, and no part of it is green. In addition, its flower is a thick stalk crowded with corpse-white blossoms that emerges directly from the ground and reaches no more than a foot high. Once the flowers are pollinated by flies and bees, they are replaced by round, white fruits that make the flower stalk look like an ear of corn. Bears and deer eat the fruits and disperse the seeds.

Once those seeds are dispersed, things really get weird. Bear corn's many seeds are minuscule—somewhat reminiscent of the fungal spores that also happen to be plentiful in pine and oak woods at this time of year. (Case in point: a previous skidding halt on the bikes had been caused by a big patch of apricot-colored chanterelle mushrooms. Near the chanterelles we found a pure white *Russula* mushroom. The first were gathered for dinner, the second was thrown against a tree trunk to watch its Styrofoam-like texture explode.)

As with fungi, the bear corn's seeds disperse easily and can stay dormant but viable in the soil for about 10 years. This gives them the best chance to accomplish what seems like an impossible task. The tiny seeds have almost no energy reserves and no means of producing food for themselves (that would require chlorophyll). If a seed is to succeed, it must find food immediately upon germinating or it dies. The food it seeks is sugar stolen from an oak tree.

The seed delays germination until it senses chemicals exuded from oak roots. The chemicals don't travel far through the soil, so if a seed gets the go-ahead, salvation must be near. A baby root called a radicle grows toward the signal, guided by chemotrophic attraction. Once contact is made with an oak rootlet, the bear corn's own tiny root immediately starts sucking sugars from its host.

Using sugars from the oak's photosynthesis, the bear corn forms a nodule on the root. The nodule grows and multiplies, gaining strength until it can support a flower structure and send its own miniscule minions into the world. Then the fruiting stalk withers into the weird, crumbling, black patches we'd just encountered.

This parasite doesn't seem to have a noticeable negative impact on its host, but why would an oak allow a freeloader to steal its hard-earned sugar? The answer may lie with those mushrooms spotted earlier. Oaks readily form beneficial mycorrhizal relationships with dozens of species of fungi. Fungal mycelia provide additional nutrients and water that help oaks survive in dry, sandy soils. In return, the oak shares its yield of sugars with the fungus. Perhaps the bear corn's seeds duplicate a fungus's flirtatious attempts to initiate a relationship, and the oak lets down its guard.

From sandy glacial sediment grow drought-tolerant trees that are assisted by beneficial fungi and robbed by sneaky imitators. As we ride off down the trail, it's clear that this esker's sinuous path connects more than just points on a map.

In Which I Join a Potter Wasp for Lunch

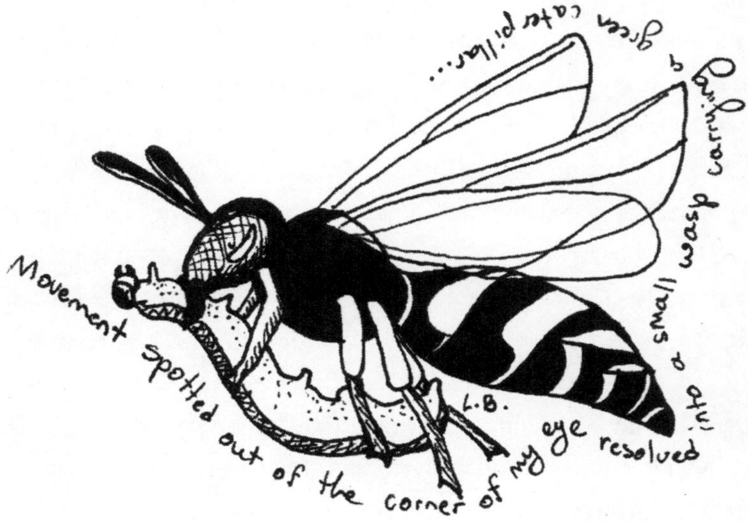

Laura Berlage

My friend and I were grateful for the light breeze that sent our patio umbrella flapping, and also for the shade of said umbrella. Somehow we'd chosen what may be the last hot day of summer to go on our annual bike adventure. Eighty degrees and humid over the course of almost 20 miles was sufficient to get the sweat dripping.

It was worth it, though, to swoop through the hilly backroads of the Chequamegon-Nicolet National Forest and emerge from the woods at the Farmstead Creamery Café and Market on North Star Homestead Farms. Run by a family of strong and talented women, this diversified network of pasture-raised poultry, sheep, and hogs, as well as raised-bed gardens, a high-tech aquaponics system, and permaculture practices, draws lots of hungry admirers.

An outside table with a big umbrella was the perfect place to relax and wait for our local, organic, home-cooked, delicious lunch to arrive. And then my day got even better.

The movement I spotted out of the corner of my eye resolved into a small wasp carrying a green caterpillar. The caterpillar was about the size and diameter of the wasp's own body (one-half-inch by one-eighth-inch or so), but neither seemed to be struggling. The wasp hovered and wandered for a bit in the vicinity of the umbrella's crank arm. Finally she landed and started the slow process of dragging the limp larva into a mud-caked hole in the end of the crank's handle.

This observation tripled my excitement for the day. While watching the drama, one of the farmers popped out of the café carrying our plates of food. Naturally, I waylaid him at the table and made him check out the action. I'm not sure he knew what to make of it—or my enthusiasm—at first, but I think I convinced him to be impressed before he went back inside.

After a few minutes the last segment of green disappeared. Then, suddenly, the wasp backed out of the hole and flew off!

Over the course of lunch, I had to put down my delicious pesto-melt sandwich several times to capture photos of the wasp landing and of her black-and-yellow-striped abdomen circling around inside the entrance to the little mud and metal nest. She brought a load of something each time, but never another caterpillar while we watched. The bundles clasped in her forelegs were dark and amorphous, impossible to identify without robbing her of her quarry.

After our amazing dessert of sheep's milk gelato (blueberry fudge flavor—yum!), we started the long, hot ride home.

Using an amazing website called Bugguide.net, which is hosted by the Iowa State University Department of Entomology, I posted my blurry cell-phone photos for identification help. Within hours a contributing editor with experience in wasps replied with a name: subfamily Eumeninae.

Now, understandably, my fuzzy photos didn't give enough information to identify this wasp to species. Even so, just knowing the subfamily she's in gave me some clues to her lifestyle. Also known as potter wasps, this diverse group of solitary wasps uses mud to create a variety of nest shapes. Some potter wasps make round-bottomed, narrow-necked nests that look exactly like traditional pots. Legend has it that some Native Americans based their pottery designs on these nests.

Other potter wasps make their nests in preexisting cavities, like the end of a crank handle on a patio umbrella or any other hollow or crevice they can find. Also called mason wasps, they use mud or sand for construction.

Inside each nest chamber she constructs, the female wasp lays one egg and then goes about the business of storing up baby food. Adult potter wasps eat nectar, but for their growing offspring, the mothers collect moth larvae like the tiny green caterpillar I witnessed going to its death. Beetle larvae and spiders may meet a similar demise. In any case, the wasp delivers a paralyzing sting to the prey, but the toxin does not kill them. Dead food would soon decompose. These stay fresh but immobile, ready for the hungry wasp larva to feast on when they hatch.

It may take anywhere from a few weeks to a year for the larva to pupate, metamorphose, and mature. Then, with powerful jaws, the adult wasp breaks open its earthen nursery and emerges to find a mate.

While watching this black and yellow lady go about her business, two main thoughts swirled through my head. The first was the importance of a healthy insect population on an organic farm. Using pesticides would have killed off this petite predator, removing a free source of caterpillar control. North Star Homestead Farms is doing pest management right. The second thought was about how tightly packed this world is with *life*. Every nook, every cranny, every surface is someone's home. Experiences like this make me glad that it's my home too.

Isle Royale: How Did You Get Here?

Clare Nieds, age 12 (squirrel), Sherry Nieds (moose)

My family and I heaved our backpacks, loaded with high-energy foods, waterproof tents, and warm clothing, onto the ferry dock as the gray dawn was beginning to break. Waves crashed around the *Voyageur II*, and gusts of damp wind threatened to steal our hats. The ferry captain came out to greet us, passenger list in hand. The whole blurry-eyed group expected him to say something like "All aboard!" Instead, with a bemused smile, he projected his voice over the bluster to the waiting crowd and said, "We're not going."

At first there was stunned silence, with a few hesitant

chuckles, since he looked like he might appreciate a good joke. But he wasn't joking, and neither was Lake Superior. Eight-foot waves and 30 miles-per-hour gusts across a 20-mile expanse of cold open water are nothing to mess with. "This is your one chance for a refund," he said, "otherwise be here by 4:45 a.m. tomorrow."

So, for Plan B we headed up to Grand Portage National Monument Visitor Center. The park ranger at the front desk didn't miss a beat when we told him of our delayed departure to Isle Royale. "Welcome to a long tradition of people waiting at Grand Portage for good enough weather to begin a journey on Lake Superior."

The voyageurs who rendezvoused here every summer through the height of the fur trade era called Lake Superior "The Lady." And this lady makes her own rules. Icy water, big winds, and craggy rocks don't make for safe or easy travel. Gardens of shipwrecks can attest to that. But rising out of the crystal clear water 14 to 20 miles from shore is a bit of an anomaly. Isle Royale, a 45-mile long and 9-mile wide bedrock island, is teeming with life that somehow made the treacherous journey.

The next morning, with the rough and rolling, cold and wet ferry ride behind us, we disembarked gratefully at the Windigo dock on the southwest corner of the island, joining the many lives already there. After cooking oatmeal in the campground, we hoisted our packs and started off down the trail.

Before long we met several pairs of hikers just ending their trips. We asked about their hometown, which ferry they took, and their route on the island. In essence, we were asking "How did you get here?" Mostly they used the water route, but one couple arrived by air in a float plane. Historically, making winter crossings by dogsled was also common. Isle Royale is not an easy place to get to, or to get around, and yet life surrounded us on all sides. Soon I started asking "How did you get here?" to everything we saw.

While relatively few humans arrive on the island by air, many of the island's wild residents and visitors arrived that way. The haunting wails of loons drifted up from every lake we passed. Chickadees chattered above us as we hiked, while flocks of cedar waxwings played follow the leader between berry-laden mountain ash trees. It's not hard to imagine how the birds got here, or how the seeds of their favorite fruits flew here either.

Fruit-bearing shrubs like chokecherry, mountain ash, and serviceberry are often pioneering species, since their seeds are air-dropped in a packet of fertilizer. The vibrant red fruits of thimbleberry caught our attention too and supplemented our quick breakfast. Slowing down to pick berries left us open to attack, though, from the delicate mosquitoes who once made the dangerous crossing too.

While the whine of mosquitoes was pleasantly rare during the late-summer days, red squirrels scolded us incessantly. A water route seems like the only plausible explanation for the squirrels' presence. Did the first squirrel on the island drift here on a fallen tree or raft of vegetation? What a frightening ride without a motor, rudder, rain gear, warm sweater, or pack full of food! How many attempted journeys failed—with no refund? Now the red squirrels have been here so long—separated from the mainland population by that arduous journey—that they are considered their own subspecies: *Tamiasciurus hudsonicus regalis*.

Not just a scolding but a crashing in the brush caught our attention. Through the dense spruce trunks we caught a glimpse of the hulking silhouette of a cow moose as she vanished into woods. Moose are thought to have swum here from Canada around 1900 during a time of overpopulation on the mainland. And they didn't have a bowl of warm oatmeal to greet them at the end. The trees they browse were here already, though, having colonized the island (by air or water?) starting shortly after the glaciers retreated 11,000 years ago.

Having sighted one of the iconic species of Isle Royale, we were now on the lookout for the other: wolves. Large, hairy scat on the trail signaled their presence, as did a few big four-toed tracks among the boot prints on muddy trails. With only two or three wolves left on the island as of autumn 2015, that sign was more than I'd dared to hope for. Wolves likely crossed the ice bridge in the winter of 1948–49, and they helped stabilize the moose population for many years. Now wolves are suffering from inbreeding—one liability of living on an island—and the National Park Service is exploring management options for both wolves and moose.

Air, water, ice. It is almost easier to imagine how the

1.1 billion-year-old bedrock of this island came to be than to imagine how such a diversity of seemingly fragile organisms came to colonize it. Perhaps most telling is that not everything made it. Bears, raccoons, skunks, porcupines, cottontails, and snapping turtles are notably absent. For humans too, Isle Royale National Park is one of the least-visited parks in the country, and one of the most costly to visit (measured in time, money, and the discomfort of seasickness). Perhaps I'm biased, but I think that makes the creatures who do get here a little more special.

Isle Royale: How Did Orchids Get Here?

Leslie Strapon

"Rattlesnake!" I cried out as I stopped abruptly and shifted my pack so I could look down at the ground. Excitement (not fear) was surely evident in my voice, but not many people would recognize the source of that excitement. No snake, but several whorls of grayish-green leaves with white markings hid within the leaf litter along the Minong Ridge Trail on Isle Royale. The checkerboard pattern that adorns the leaves is said to look like a rattlesnake's mottled camouflage, while the broadly oval leaves are similar in shape to plantain leaves (a common plant of yards and disturbed areas). Together, these features resulted in the name rattlesnake plantain orchid for this beautiful little plant.

Throughout my hike across Isle Royale, I continued to see clusters of the three different species of rattlesnake plantain orchid along the trail, and also several other less distinctive orchids, already past flowering. The family Orchidaceae is one of the two most diverse plant families (Asteraceae is the other) and contains more than 22,000 species across the globe. The flowers range from extravagantly beautiful to dinky and unobtrusive. Many orchids have evolved highly specialized pollination systems. Some flowers look just like female bees and attract amorous male bees to do their cross-pollinating. Others have special landing pads on their petals and distinctive scents wafting out to entice visitors.

What fascinates me the most, though, is what happens after an orchid is pollinated. The seeds that form are almost microscopic, and more than a million of them can fit into the capsule that develops from a single flower. Being so tiny, they lack an endosperm. That little packet of starches, oils, and proteins usually gives seeds the energy they need to germinate, grow their first leaves and roots, and get along until they can sustain themselves. Corn, beans, and squash are great examples of seeds with large energy reserves, and their seeds sustain us too.

Orchids have found another way.

For a dustlike orchid seed to germinate it must first be infected by a specific fungus. Not all orchids form a symbiosis (a close living relationship between two species, whether or not they both benefit) with the same fungus. Each orchid has its own one or two fungal species, or maybe a few, that will work for it.

Fungal mycelia provide the seed with sugar and nutrients in place of the endosperm. It is unclear if the fungus receives anything in return. The seeds can't germinate without their fungus, except in lab conditions with a source of sugar. Once colonized a baby orchid plant, called a protocorm, grows and eventually produces leaves and roots. Some orchids never produce leaves or chlorophyll at all and live out their days entirely parasitic on a fungus.

Because I know these basics of orchid germination, the question I'd been asking everyone and everything on the island baffled me even more: "How did you get here?" How could these tiny, fragile seeds make such a long and treacherous journey 14 to 20 miles across the lake, and then happen to land where their friendly fungal partner was already established?

My amazement only deepened when I used the latrine near our Rock Harbor campsite for the first time. The educational poster on the inside of the door bragged that Isle Royale is home to 32 species of orchids. This amazing coincidence happened 32 times—at least—over just several thousand years. Wow.

When I shared my amazement with a mycologist friend, though, he was less impressed. "It's almost inevitable," he shrugged. As it turns out, the numerous, minuscule seeds of orchids are perfectly adapted for wind dispersal—in much the same way that pine pollen drifts on the breeze. That's how orchids colonize tree trunks in the rainforest. Likewise, fungal spores are often wind-dispersed, even over many miles, and the species of fungi that orchids parasitize are quite common.

A little more research revealed that Ontario—the province upwind of Isle Royale—has more than 50 species of orchids on its mainland. Maybe 32 species on Isle Royale isn't that impressive after all.

With this new information, my amazement at the simple fact that orchids have colonized on Isle Royale shifted to admiration for the adaptations that brought them to the island in the first place, and for the intricacies of their symbiotic relationships. Perhaps my question "How did you get here?" should refer not to the island but to this moment in time and the state of exquisite adaptation housed in a ground-hugging, snakeskin-patterned cluster of grayish-green leaves.

FALL

Packing for a Trip

Gavin Porter, age 10

Mind racing, I drove through the crisp, clear late-summer sunshine. Red maples flamed in the swamps, and the ditches were golden with sunflowers and senescing bracken ferns. But I didn't notice them. All morning I'd been darting around the house like crazy, packing for a trip to the Boundary Waters. Now, I was on my way into town for one last round of forgotten things.

A sudden flash of black and white snapped me into the present. A medium-sized bird with long, pointed wings and a short, compact body darted above the road. One white stripe across each dark wrist flashed as it dove and swooped erratically. Soon a second, a third, and several more acrobatic birds appeared in the clear blue airspace above trees, homes, and the Namekagon River.

The diet of common nighthawks consists almost entirely of flying insects. There's a good reason their fall migration has started just when I'm embarking on a wilderness expedition. As the bugs die off, canoe camping becomes more pleasant and the nighthawks must move on. In addition, nighthawks (and humans) are unable to go into torpor on chilly nights, so must carefully avoid temperatures too cold for their respective down coverings.

As the shortening days and crisp nights trigger nighthawks and other birds to start their migrations, this weather triggers my itch to launch a canoe onto some wild chain of lakes. I love preparing for canoe trips in the Boundary Waters of northern Minnesota. I weigh out the portions of dried pea soup mix (four ounces per person), measure out the macaroni noodles (one cup per person), and thoughtfully pack a little bottle of hand sanitizer in the bag with each roll of toilet paper. Once I'm on trail, a few handfuls of late-ripening raspberries or blueberries or a steaming cup of sweet gale tea will supplement the dried foods.

Nighthawks use a similar strategy. Like most birds, they build up stores of fat to fuel their trip, but they also stop and snack on local foods along the way. Their beaks look tiny, and hardly protrude beyond their heads, but open into cavernous mouths well suited to scooping insects up in midair during their buoyant, spurtive flight. Nighthawks are attracted to clouds of insects that gather around streetlights at dawn and dusk, but seem to hunt by sight and can't forage after dark. Only because of migration did I get to see this flock in broad daylight, and so close to the ground.

For several days I'd been making my own spurtive flights up and down the stairs, to the kitchen, to my bedroom, and back to the basement, rounding up gear and checking the list twice. Seeing these birds gave me some much-needed perspective. While my trip will last only four days and cover 400 miles of driving plus 20 miles paddling, nighthawks have one of the longest migration routes of all North American birds. From northern Canada, they travel 2,000 to 6,000 miles south to their winter range in Argentina.

I've started to look forward to seeing their migration as I'm coming or going from a wilderness trip. Last year at about this time I saw clouds of nighthawks migrating along Lake Superior as I drove back from Grand Marais. This year the migration watchers at Hawk Ridge in Duluth already counted 4,140 nighthawks on August 14, and 2,020 more on August 15!

These birds will arrive on their wintering grounds in October, spend a few months, and by the end of February, head north again. Nighthawks are one of the latest arriving spring migrants, due again to their intolerance of cold and need for flying insects.

Last June, students on the Wisconsin Master Naturalist course spotted a nesting nighthawk during a banding expedition in the Moquah Barrens. The bird flushed or we never would have seen it. Two speckled eggs in a scrape—

no nest—lay revealed on the barren, recently burned earth. Mama sat nearby, camouflaged perfectly as she perched along a branch surrounded by ferns. Just 50 days after hatching, those chicks must be full grown and ready to take their first trip south.

Tomorrow morning at dawn a van full of wilderness travelers will head north. In the gray morning light, we'll be crossing paths with flocks of nighthawks flying south. Safe travels to all.

Bugling at Dawn

Leslie Strapon

The sky was beginning to lighten into grays and pinks as the group gathered at the Clam Lake Guard Station to meet Laine Stowell, elk biologist with the Wisconsin DNR. Slowly rolling down Highway 77 in our car caravan, we barely noticed that dawn was breaking.

The first stop was behind a storage facility at the end of a paved road with several small homes. Riding in the truck with Laine, I'd been privy to the steady beeping that signaled a cow with a radio collar was nearby. Driving with his left hand, Laine's right hand gripped the pole of the big receiver antenna punched through his roof and spun it slowly. As the antenna pointed in the direction of the cow's VHF collar, the beeping grew louder. She was close, he said, between the side road and the highway.

I once wrote about how the presence of elk in the Chequamegon-Nicolet National Forest made it feel wild. On that morning the group had hiked a mile or so into the woods on a dirt road. Today's suburban adventure didn't convey the same mystique.

Parking at the end of the lane, we all tiptoed out of our vehicles and gently pushed the doors closed, minimizing sound as much as possible. Gathering around Laine, we listened as he put the small cow call to his lips and squealed out several expressive notes. When that failed to elicit a response, he continued with the longer bull bugling device. The flexible tube was covered in camouflage with shiny black pieces at both ends. Putting his lips to the mouthpiece, Laine let out a high-pitched wail. We hoped another bull would hear the bugle and take up the challenge.

We stood expectantly: ears open, breathing controlled, arms wrapped against the morning chill. We admired the beauty of the morning light and dappling of fall colors. Silence. The elk were close, but not cooperative.

Elk were extirpated from Wisconsin in the 1800s due to overhunting and a rapid decline in habitat. In 1995, 25 elk were released into the national forest near Clam Lake, and the DNR now estimates the population to be between 160 and 170 animals. Several dozen more elk are scheduled to arrive from a Kentucky herd in March 2017. The new elk will boost the genetic diversity of this herd and increase the population to the point where a limited hunting season can be considered.

Piling back into our vehicles, we caravanned to a few more sites, each deeper into the woods than the last. Laine had woken up early to scout the area and had located several cow-calf groups in the vicinity. During the rut—which begins around August 25 and can last into the middle of winter—where you find cows, you also find bulls. They were there . . . We could hear the beeps from their radio collars . . . but none made a peep in response to Laine's calling.

As the sun rose higher we bumped still deeper into the forest. A long, narrow clearing appeared and we pulled off to the side. This was the ELF line, and its open, grassy habitat is one of the main reasons that this area was chosen for elk reintroduction in 1995. ELF was a US Navy project that used extremely low frequency (ELF) radio waves to communicate with deeply submerged submarines. The transmitter operated from 1989 to 2004 and consisted of two fourteen-mile transmission line antennas in the shape of a cross, with the transmitter station at their intersection. The lines were removed in 2008. Its legacy is that some of the clearing that was once mowed for maintenance is now kept open for wildlife habitat. The elk like the freedom of movement and the tender new growth that the cut area provides.

We walked several yards uphill, away from the cars and

the valley of the Torch River. By now the group was getting a bit restless; still hopeful, but also resigned to the possibility that we might not hear any elk.

Laine made another cow call, and almost immediately the haunting bugle of a bull echoed in the distance. Did you hear it? we asked as we looked around the group in excitement. Not everyone had heard. Laine called again. The distant elk bugled again, and then, after a pause, another bugle sounded closer and from a slightly different direction. We grinned and pointed in the direction of the sounds.

At least one bull was pretty close, and in the opposite direction from which we'd been walking. Back down past the vehicles we ambled through the ELF clearing and stopped on a knoll above the river. Laine and the two bulls called back and forth several more times. We even saw a flash of warm tan hair through the trees, likely an elk feeding on the other side of the river.

Standing there in land cleared by the military, admiring the cast of sunlight, and listening to the uninhibited sound of animals going about their essential business of mating, I was struck by the contrast between wild and human contrived. The elk were extirpated because of humans. They returned because of humans. We found them by using very high frequency (VHF) radio collars in an area cleared for extremely low frequency (ELF) communication—both manmade. And yet the elk paid almost no attention to us. Their instincts, their drive to mate and survive, are the same as ever.

The wildness of managed populations is a philosophical question that isn't resolved in my own head any more than it is decided among scientists or the public. Humans have complicated things, as usual. But still I'm grateful for the opportunity to listen to the majestic bugle of an elk on a crisp fall morning just minutes from my home.

Discovering Pin Lichens

Zoe Braun, age 7

I was standing at the information desk in the foyer of Lakewoods Resort near Cable, Wisconsin, when the lichenologist returned from the field. Matt Nelsen is a postdoctoral researcher at the Field Museum in Chicago, and he gave a talk on lichen biology during the Northwoods Foray of the North American Mycological Association.

I'd missed his talk due to my coordinator duties, so I eavesdropped eagerly as he chatted with a colleague about a chunk of dead wood he was holding. He'd been out exploring the Rainbow Lake Wilderness Area northwest of Drummond, where in 1990 the US Forest Service had found 190 species of lichens during a survey.

According to Matt, he'd just happened to notice a decorticated snag that looked like it would be prime habitat for pin lichens. He said this with a lopsided smile and self-deprecating chuckle. Yes, scientists are aware that many of the words they use could be replaced with plainer language. *Decorticated* simply means the bark, rind, or husk has been removed. In the dictionary, "decorticated peanuts" is given as the example. In Matt's defense, texts on lichens use the word *decorticated* frequently and, in my opinion, big words can often add poetry, precision, and entertainment value to a sentence.

Anyway, Matt was right. The decorticated snag was covered in pin lichens. When he held out his specimen of fibrous wood he'd sliced off so that we could get a better look, all I could see was a little black stubble on the surface.

That was enough to pique my curiosity, though, and I followed Matt downstairs to the voucher room, where scientists worked with microscopes, technical keys, and each other to identify the more than 470 types of fungi collected at the foray. Although lichens are a symbiotic relationship between a fungus and an alga, they are classified by their fungus, and so their study traditionally falls under the jurisdiction of mycologists.

One of those mycologists had brought with them the most amazing dissecting microscope I've ever seen. This type of microscope is meant for viewing objects through relatively low magnification, as opposed to a compound microscope that uses higher power magnification to look at things like bacteria and cells. The paired eyepieces provide for three-dimensional viewing, and lights all around the microscope's stage illuminate the subject perfectly.

On the way downstairs, Matt had rattled off a bunch of the pin lichen's unique characteristics. Some species of pin lichens are used as old-growth indicators, because they are picky about their habitat being undisturbed. The thallus, or body of the lichen, is inconspicuously buried in the wood. Each of the pin lichen's "pins" is a stalked apothecium—a type of reproductive structure. At the tip of each stalk is a capitulum—a round cluster of dark brown spores enclosed in a thin membrane. Unlike many fungi who actively expel their spores, the outer coating of a pin lichen's capitulum simply dissolves, and the spores might cling to the fur or feathers of passing animals or be dispersed by wind or rain. Underneath the capitulum is a ring of yellow-green crystals containing vulpinic acid, and whose purpose is unknown.

After Matt placed the drab chunk of wood on the microscope's stage, it took me a few minutes to adjust the eyepieces and focus. As a naturalist, most of my training has been in looking at the world from the thousand foot view, with the naked eye, or with a hand lens. This was a whole new world. Linear ravines and ridges ran the length of the pale, fibrous substrate. On those ridges grew a savanna of Dr. Seuss–like, stalked apothecia "trees." And decorating the lower canopy of each was a ring of luminous green crystals. I felt like I'd discovered a forest with foliage made of gemstones.

Up and down I focused, immersing myself in this fairyland. My eyes galloped over the ridges, alighted in the canopies, and shimmied up and down the slender-stalked apothecia. How does something as beautiful as this exist in our world with so little notice taken of it? Any lichenologists reading this are probably chuckling at my naivety, since they learned about it long ago. A portion of the rest of you probably think I'm crazy, and maybe a few others are grabbing their boots, hand lenses, and knives to go investigate the nearest decorticated tree. Looking for lichens in Wisconsin should be pretty easy—there are more than 662 species recorded from across the state, and 255 of those are found right here in Bayfield County.

So, filled with wonder and my new knowledge, I wanted to see if I could discover pin lichens again, this time on my own and in their natural habitat. Remembering Matt's words, I headed down the street in search of a decorticated tree. At the edge of the forest in the Cable Natural History Museum's natural play area stood an old, bare white pine that I'd never paid much heed. Nosing in, I found them right away. Tiny black pins covered at least three square feet of the snag's sunny side.

Each new natural neighbor I discover becomes a friend whom I can meet up with during a day in the woods.

Fun Guys in the Fen

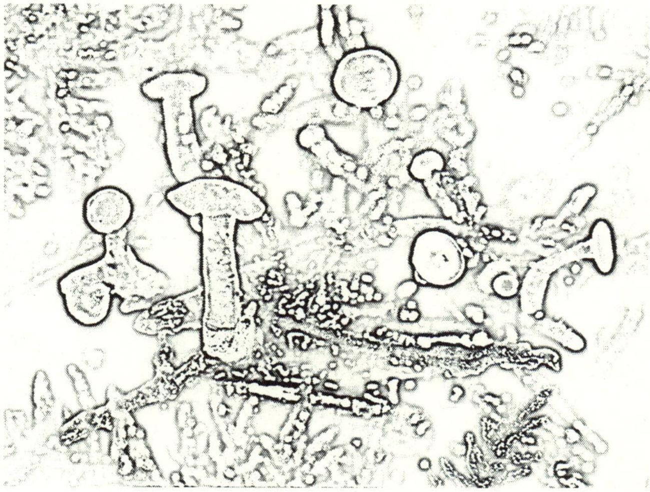

Modified photo from Michael Beug

None of us had rubber boots handy, so when we came to the log bridge across the moat we took off our shoes and went barefoot. I enjoyed the firm roughness of the decorticated log on my soles, and then the cool squish of the sphagnum moss in the floating mat as I stepped into the magical world of the fen.

I love bringing appreciative people into the Namekagon Fen, and my companions on this day were wonderful. Peter Kennedy is an associate professor at the University of Minnesota, and he had just given a talk called "Knee Deep in Peatland Fungi: Mycorrhizal Community Diversity and Composition in Fens and Bogs of the Northwoods." He'd already been to one nearby bog as part of the North American Mycological Association's annual foray, held in

2017 at Lakewoods Resort near Cable, Wisconsin, but when I offered to show him a giant fen too, his eyes lit up.

As I gave some final instructions to my volunteers at the information desk and prepared to play hooky from the foray for a couple of hours, Michael Beug walked up with his camera and began chatting with Peter. Michael is a professor at Evergreen State College in Olympia, Washington. Back in the day, he teamed with the college's ornithologist to study the impact of DDT on calcium transport in birds, and this work played a major role in getting the pesticide banned. He and Peter had met during Peter's undergraduate studies. Soon Michael went off to grab his camera so he could join us.

So there I was, ankle deep in cold, wet sphagnum with two PhDs who were talking a mile a minute about shared experiences and recent news, with occasional breaks to point out neat things in the fen.

Waving his hand over a patch of dead sphagnum moss, which looked bleached and scraggly compared to the velvety-red living moss beside it, Michael informed me that it had been killed by a fungus. He poked around for a bit, hoping to find the delicate, pink mini-mushroom forms of *Roseodiscus sp.* No luck.

I offered up the impressive fact that my previous explorations had revealed this fen to be more than 40 feet deep, with the bottom still out of reach of our current tools. We also found pitcher plant leaves, a tiny frog, a few late-blooming bog laurel flowers, a couple of red-orange waxy cap mushrooms, and a tiny white pine seedling growing out of a sphagnum hummock.

Without hesitation Peter dove elbow deep in the moss, following the white pine's stem down to its roots and out to the fine roots to look for mycelia. Tree roots aren't that great at directly acquiring the water and nutrients they need. Most but not all plants have instead cultivated a symbiotic relationship with fungi in the soil. A web of threadlike fungal mycelia connects with the tree's roots to bring it water, minerals, nitrogen, phosphorus, and even protection from pathogens. In return, the tree feeds the fungi sugars it produces through photosynthesis. As Peter had described in his talk, this mycorrhizal (fungus-root) relationship is important in bogs and fens because they are especially challenging low-nutrient environments.

We replanted the tree (although lack of mycelia suggested it didn't have much of a chance) and wandered some more.

We were almost back at the bridge when things got even better. I pointed out a little black spruce grove where orchids bloomed in July. We wandered in. Michael plunged his hand into the sphagnum and brought up a handful of wiry black spruce roots—their finest tips neatly capped by a white sheath of fungal mycelia. These trees were thriving with the help of mycorrhizal fungi.

Then the guys spotted some big brown mushrooms. "*Hebelomas*," Michael called them. "They're boring mushrooms nobody looks at," he told me, but his friend Henry Beker from the University of London had just finished the monograph on the European *Hebelomas*, so

Michael wanted to help him document the ones here too. They may be mycorrhizal, which makes them interesting to me.

As Michael was stooped over his tripod getting the drab, brown mushrooms in focus, Peter happened to see something else between his feet. I couldn't even begin to understand what they were exclaiming over until Peter brought up a small clump of sphagnum moss for me to see. Clusters of miniature mushrooms bejeweled the mess, and their white bodies glowed in the sunlight like frosted glass ornaments.

Neocudoniella radicella: this little mushroom is found in sphagnum bogs where black spruce rootlets turn up out of the soggy mat to get air. It was a "lifer" for Michael, a mushroom he'd never seen before. "The fact that I knew it existed in a place that is incredibly beautiful and magical made it an object of desire," he later told me.

A little handful of the gems went into Peter's collecting basket and eventually made its way through the vouchering process in the basement of Lakewoods Resort. One last hurrah came that evening during announcements. Our tiny gems had been chosen by the mycologists as the find of the day!

That was nice, of course, but there was nowhere to write on the voucher form that the fen where these mushrooms grew was filled with bare toes squishing the sphagnum, joy bubbling up in the sunshine, old friends reconnecting their bonds, and new friends sharing an adventure. The fen is filled with treasure!

Bryozoans

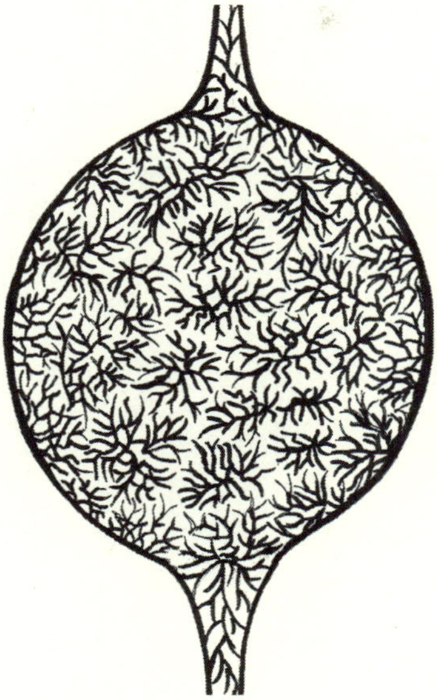

Sara Balbin

The solid yet gelatinous blob felt pleasantly slimy in my hand when the girl handed it to me. Although we stood only waist deep in the Namekagon River below its headwaters dam, neoprene wetsuits protected us from the surprising cold of late August water in 2015. When we showed the patterned blob to scientist Toben Lafrancois, he recognized it immediately: bryozoan. It was a colony of tiny invertebrates all clinging to the stem of an aquatic plant.

The firm-Jell-O-with-a-textured-surface feel of the bryozoan triggered my "gross" reflex as much as my "cool" instinct. I was impressed when the group of teenage girls I was snorkeling with didn't bat an eye. Showing interest (not too much) but not disgust, they all had a look before we put the colony back in the river.

I suppose I shouldn't have been surprised, because this group had been snorkeling and studying river ecology with Toben and Ian Karl, the experiential program coordinator, all summer as part of Northwest Passage's "New Light Under the Surface" program. Through a National Park Service Youth Partnership Program grant, these at-risk kids had been armed with waterproof cameras, wetsuits, and snorkel gear, and were trailblazing a realm experienced by incredibly few, all while developing healthy life habits.

My afternoon with the girls was incredible. I'd never seen that side of the river before, and I enjoyed being the student as they helped me learn to snorkel, use an underwater camera, and follow safety protocols. Memories of that day have stuck with me, with the lumpy mass of the bryozoan rising to the top as a sort of touchstone.

So this fall, when I received several photos of bryozoans with identification requests, memories of that snorkeling experience bubbled up, and I decided it was time to investigate further. Local artist Sarah Balbin kindly led me out to a patch she'd found on the narrows of Picture Lake near Cable, Wisconsin, and I collected a small specimen to examine later under the Museum's new digital microscope.

With my sleeve rolled up, I plunged my arm into the cool water and tugged on the plant stem bisecting the bryozoan. The dead, decomposing stem released much more easily than I anticipated, and soon I was looking at a pair of miniature football-shaped colonies on the deck of my kayak.

While the blob looked more like a patterned glass art piece than a living being, it is actually formed of hundreds of genetically identical, physiologically interconnected animals called zooids. Each individual zooid is made up of an outer sleeve-like structure, a mass of organs that can move within the sleeve, and the lophophore, which is a ring of tentacles that filters food particles out of the water. Altogether, a zooid may be only half a millimeter long, so none of this was visible to my unaided eye.

While each zooid is capable of independent feeding, digestion, and reproduction, they share certain fluids and tissues that unify the colony, and it's impossible to separate them entirely.

Back at the Museum, I examined the bryozoan under our nifty new microscope. The brown and yellow pointillist designs I'd admired in the field resolved into a conglomeration of various-colored polka dots, often with concentric rings. Nothing looked remotely like the feeding tentacles I'd read about. I snapped some photos through the scope and sent them off to Dr. Lafrancois for clarification.

"They are statoblasts," he wrote back, "or dormant buds getting ready for winter." While bryozoans can reproduce sexually, they also produce these statoblasts, which are sort of like tiny survival pods that develop while attached to zooids. Inside each bivalve-like chitin shell hides a protozooid with some stored food.

The statoblasts of some species contain a gas bubble that will float them up to the surface. Others drift on a current or sink to the bottom. Apparently, the various shades of brown and yellow represent various stages of statoblast development, and their prevalence is a result of the bryozoan's autumnal push to create clones who can survive the winter.

The bryozoan colony will die as temperatures drop, but statoblasts are incredibly durable and can survive long periods of dormancy, freezing, dehydration, and even transport in the gut of a duck or the ballast of a ship. In the spring the protozooid will grow, bud off more zooids like itself, and form a new colony. By fall, colonies have reached their largest size, and people notice them more frequently . . . prompting them to send photos to me, and me to finally write about this incredible creature.

As I scooped the bryozoan off the microscope stage and back into its plastic bag, the texture of it in my hand brought back memories of snorkeling with those girls in the Namekagon River. We were all waist deep in discovery on a sunny August afternoon.

Josephine Lafrancois, age 7

How to Make a Beaver Sparkle

Jennifer Rutten

Cool water swirled around my rubber boots as I waded out into the dark, star-studded lake. A sense of peace began seeping in with the chill. Then, *splash!* From out of the darkness came the unmistakable sound of a beaver slapping its tail in alarm. Of course I looked up, and my light caught his swimming form in the beam.

I chuckled at my own surprise. Then immediately I began thinking of how I would tell this story to the kindergarteners during my MuseumMobile program the next morning as we passed around the taxidermied beaver's tail.

Beavers are not my favorite animal. Their drab, oily fur, plodding manner, and lumpy design don't inspire the same feelings of wonder in my heart as a cheery little chickadee, rangy wolf, or glittering dragonfly. But over the years I've come to enjoy teaching about them nonetheless.

During my internship at Acadia National Park, the lesson plan for a fourth grade field trip called "Animals of Acadia" included beavers. In big, yellow buses we started at the Precipice—the nesting habitat of peregrine falcons. On a good day, we could see their elegant forms darting gracefully in front of a craggy cliff. It took me a while to understand how peregrines and beavers fit into the same program—the only time beavers look elegant is after they've been skinned and felted into hat—but I came to appreciate their parallel histories of exploitation and steep decline, followed by protection and recovery.

One goal of the field trip was to teach the students about animal adaptations. While you can hardly find two animals who are more different, beavers and peregrines share the reputation of being extremely well adapted to their particular lifestyles.

Back on the shore of Lake Namakagon the beam from my headlamp followed the beaver on his journey, and his eyes sparkled back at me. I paused to admire the effectiveness of his oddly shaped head. Just that morning I had pulled a skull out of the education tub to show some second graders how a beaver's eyes, ears, and nose are all crowded toward the top of his head. This allows the beaver to hear danger, see where he's going, and breathe continuously, even while having most of his head and body stealthily submerged.

I also enjoyed watching the second graders react to the news that beavers have a third translucent eyelid that closes sideways and acts like swim goggles. They were jealous! I could see their little minds churn as they imagined what they could do in their favorite lake with built-in goggles. I wouldn't mind having the beaver's ear, nose, and throat flaps to keep water out too.

The beaver's rust-colored teeth caught the students' attention and spurred questions—which is one reason I love having access to dead animal parts for teaching. The orange surface isn't due to poor dental hygiene. (Although saying that might encourage kids to brush their teeth more!) The orange color comes from iron in the enamel, which strengthens the surface and buffers the teeth against acid that could cause tooth decay. The iron works even better than fluoride!

Behind the orange surface, a beaver's front teeth are made of softer white dentin. As the beaver gnaws down trees, the dentin wears away at an angle behind the enamel, resulting in self-sharpening points—an innovation that would be welcome in my knife drawer!

When I teach kindergarteners about beavers, I don't go into those details. We stick to the theme of exploring nature with our senses, which means taking turns touching a beaver pelt. They get to experience the soft, dense underfur that provides insulation and the long, glossy guard hairs that help keep the beaver waterproof. At one of my favorite schools, a little class clown sprawled out on the fur, then grabbed a corner and rolled around like the beaver hide was attacking him. His classmates ignored what must have been a familiar scene, but I had to work hard not to laugh.

The kids also pass around a cloth bag concealing a beaver tail. Flip-flop, shoe, and flyswatter are some of their guesses about the mystery object. About once a year, one outdoorsy kid will recognize the tail right away. Predictably, the kids are curious and want to touch the tail again. When I ask them how the beaver uses its tail, inevitably I have to clarify that only cartoon beavers use their tails to pat mud on their lodge. Real beavers use their tails to swim, to store fat for the winter, as a kickstand while cutting trees, and of course, for slapping the water in alarm.

I began the kindergarten lesson the next morning with the story of my nocturnal beaver encounter. Before I finished, six little hands had shot into the air, eager to share their beaver encounters too. The kids' enthusiasm was a bit like my flashlight—adding a bit of a sparkle to a common creature.

Jack Pine

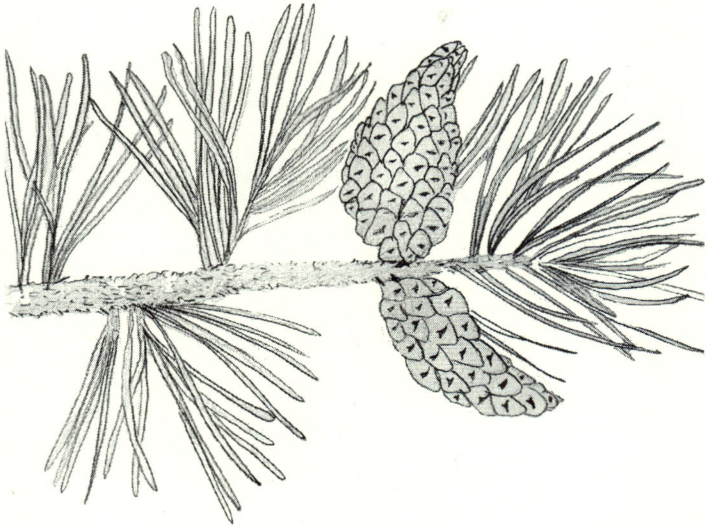

Vivianne Hanke

A fleeting sunset flamed through a ragged stand of jack pines across a bay on Lake Two. Patchy clouds provided just enough substrate to catch the color, and the faint breeze barely ruffled their reflection on the lake. Here, on a classic rock knob campsite in the Boundary Waters Canoe Area Wilderness, my friend and I perched above a scene of rugged beauty.

As it turned out, we were also perched on the border of the legendary Pagami Creek fire. First detected on August 18, 2011, the smoldering burn stayed within a bog for days. When the relative humidity plummeted to 18 percent and a north wind picked up on August 26, the fire roared to life and swept through the forest's crown. By

November it had burned 93,000 acres and sent smoke all the way to Russia. To protect the extremely popular, highly visited landscape of Lakes One and Two, the Forest Service completed a burn-out operation between the fire and those lakes.

Looking at the maps, I determined that our view was quite likely part of the burnout, and our peninsula campsite was somewhat protected by the Forest Service's efforts. The point on the north side of the bay, now glowing with the sun's last rays, was also spared. Its grove of jack pines is a respectable uniform height. While jack pines always look a little disheveled (like "something the cat dragged in," joked an old friend), at least these still had needles.

To the south, the mirrored skeletons of burned jack pines shimmered on the glassy lake, and the remains of their scraggly crowns scratched the sunset clouds at the same height as their intact neighbors. Both stands probably started their lives after a past crown fire burned through an even older jack pine forest. With jack pines, though, it might be a little harder to tell if the unburned half of the forest is more fortunate. The charred skeleton of the forest brims with life. A thick carpet of five-year-old jack pine seedlings, with a few aspen and birch mixed in, promises a bright future.

The unburned forest is getting up there in years and has attained the average maximum height for jack pines of about 60 feet. Jack pines mature quickly, and peak seed production begins at about age 20. The canopy shows signs of decay at only 75 years old. While individual trees will survive up to 200 years (the oldest known jack pine was 243 years old, found in the Boundary Waters!), there isn't much hope for a second generation of sun-loving jack pine seedlings under the stifling shade of their parents.

Jack pine is uniquely attuned to fire. Almost every burn that enters a jack pine stand will climb up the dry ladder of dead lower branches and become a stand-killing crown fire, providing seeds with full sunlight and a dose of fertilizing ash for expeditious growth. Flaky bark and resinous needles urge the fire on. In contrast, red and white pines self-prune their lower branches and grow thick, corky bark to prevent fires from crowning. They're adapted to ground fires, which clear out competition but allow old trees to thrive.

The key to jack pine's dependence on fire is its serotinous cones. Rather than referring to the sticky resin that glues the cones shut, serotiny simply describes the trait of delayed seed release. Jack pine achieves this with a satisfying series of tricks.

First, the cones are glued tightly shut with resin and may remain in storage on the tree for several decades. During the cones' extended residence, the resin protects against seed predators like crossbills and squirrels, who make short work of the nutritious seeds in the cones of other species. When a fire finally rips through the forest, the resin ignites at a low temperature, and this relatively cool flame (only 112 degrees) helps prevent serious damage. The resin is stored in reservoirs within the cone, but only burns after it has traveled down a duct to the tip of the scale and encounters oxygen there, outside the cone.

Inside, corky material on the cone scales provides insulation for the seeds. The temperature gradient created from the outside in causes the scales to curl back and open the cone like a flower. In fact, as our small campfire—composed mostly of jack pine twigs laden with tightly closed cones—crackled softly against the descending dusk, we watched in delight as the cones slowly bloomed.

Just that series of events is amazing enough, but jack pine's cones go even farther in their Rube Goldberg machine of adaptations. Although the cones open while the fire is still hot, the seeds stay safely stuck in the warm, gooey resin instead of dropping into an inferno. Only once the resin cools, shrinks, and cracks do the seeds drop out—onto fresh, habitable soil.

What's more, jack pines in fire-dominated areas produce almost entirely cones that open only with fire, while jack pines in the southern part of their range or on islands where fire is less prevalent are adapted to unfurl from the heat of the sun. Otherwise they might never open.

Sure, there is beauty in a colorful sunset flaming through the forest, but there is also beauty in the dance of fire, seeds, death, and rebirth when those flames are real.

All Flourishing Is Mutual

Zoey Sineni, age 6

Recently I began hearing weird noises on the roof. A loud plunk would be followed by a series of rattles, then silence, and sometimes a final plink. While working in the garden one afternoon I figured it out. The noises started, the short silence ensued, and an acorn launched off the roof, sailed over my head, and plinked on the roof of my car. Every slight breeze brought the same series of events hailing down on me from the large oak in the yard. The driveway soon became a roly-poly mess of round, brown nuts.

Oaks are mast trees, which means they save their energy for three to five years before producing thousands of acorns all at once. And trees don't usually mast by themselves. Somehow, all the trees in an area communicate to determine which year will be the right one. "In the old time, our elders say, the trees talked to each other," writes Robin Wall Kimmerer in her book *Braiding Sweetgrass*,[1] which won the Sigurd Olson Nature Writing Award in 2014.

Not all scientists go for that explanation. Foresters hypothesize that trees simply wait until they've built up enough energy. But that doesn't explain how trees in rich and poor habitats are ready at the same time. Mycologists now theorize that the networks of mycorrhizal fungi that connect a forest by the roots may be the agents of coordination. "A kind of Robin Hood," writes Kimmerer, "they take from the rich and give to the poor, so that all the trees arrive at the same carbon surplus at the same time. . . . Through unity, survival. All flourishing is mutual."

Why does masting together equal flourishing? It is simple math. With thousands of acorns available at once, the hordes of acorn-eating seed predators cannot possibly eat them all. Some will survive to grow new trees.

At least, that is how it works today. In the past, there was a seed predator who could clean out an oak forest, even in a mast year. Passenger pigeons, once the most numerous land birds on the continent, traveled in flocks of millions. When they descended on a grove of oaks, the twigs, branches, and even whole trees broke under the weight of their bodies. By one report, not a single acorn could be found after they moved on.

Their nesting colonies, by necessity, also followed the mast. The largest nesting area ever recorded was in central Wisconsin in 1871. The size was estimated at 850 square miles. The population was estimated at 136,000,000 birds. It takes a lot of food to support that many birds, even for the two months or less it took to mate, brood, and fledge the squabs.

Now that passenger pigeons are extinct (the last one died in 1914), it's difficult to tease out the significance of their lives, and the consequences of their demise.

What is known is the impact that extra acorns (presumably the ones no longer eaten by flocks of pigeons) have on several other species. In New York, scientists have shown that Lyme disease increases two years after an acorn mast year. The mechanism seems to be the increase in white-tailed deer numbers and local density, along with an increase in the population of mice (one of the best carriers of the Lyme-causing bacterium).

Deer help the ticks reproduce. Mice infect the ticks with Lyme disease. The ticks infect us. If the passenger pigeons were still here to eat all the nuts in one fell swoop and prevent an increase in the local concentrations of deer, mice, and infected ticks, hypothesizes ornithologist David

Blockstein, would that also keep the spread of Lyme disease in check?

The world is a complicated place, and we can never fully predict the consequences of our actions. None of the 19th century market hunters—who slaughtered thousands of passenger pigeons per day with the conviction that they would always be plentiful—would have ever suspected that their actions might result in their descendants suffering from a tick-borne disease.

Jerry Niemi, from the Natural Resources Research Institute at the University of Minnesota, Duluth, recently spoke on the rise and fall of bird populations in Minnesota and Wisconsin. Since the year 1600, he recounted, 131 species of birds have become extinct. Today, another 1,213 species are facing the threat of extinction. That is 12 percent of all known species of birds. What cascade of events will their losses trigger? What kind of avalanche will climate change bring? Certainly not one as quaint and harmless as acorns tumbling off my roof.

Kimmerer's wisdom sums it up for me: "What happens to one, happens to us all. We can starve together or feast together. All flourishing is mutual."[2]

Not Mosquitoes

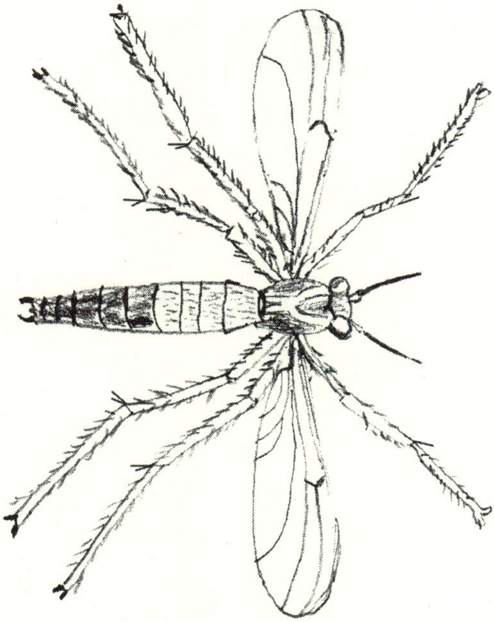

Sue Thurn

One evening a few weeks ago I looked up from my computer to see a thick swarm of mosquitoes at my window. Hundreds, maybe thousands, of the leggy little buggers bounced against the screen in the lamplight. More swarms greeted me in the morning as they knocked against the windows and formed a gauntlet to my car. Several carpooled to work with me. Others joined me while I was brushing my teeth the next evening.

The bugs were everywhere. But none had landed on me, or tried to bite, or even buzzed in my ear. That's not typical mosquito behavior! So, in the name of science, I caught one out of midair and gently squashed it. Even a hand lens couldn't tell me much about this tiny tangle of legs, but after a photo shoot with my macro setting, I was able to ascertain this critter's innocence: it was definitely not a mosquito.

The spines on its knees were my first clue. The lower joint of each leg had at least one and sometimes multiple little pointy prongs sticking out of it. The two tiny wings were clear and delicately veined. The abdomen was narrow and dark with a lighter tip. And, most odd, a projection on the underside of its thorax, what we humans would think of as our chest, almost looked like a pale-colored mite clinging like a monkey on his mom.

Stumped, I sent my photo around—first to one entomologist friend, then another. When between the two of them I came up with the family Mycetophilidae and the common name fungus gnat, I sent the photo to my mushroom guy, and he confirmed, "Definitely Mycetophilidae. Huge family over a thousand species. Google glowworms . . ."

Mycetophilids are in the order Diptera with mosquitoes, gnats, and other two-winged flies. As I parsed the Wikipedia article everything began to make sense. They are described as having a "strongly humped thorax and well-developed coxae." Did you know that an insect's coxa is the base of its leg, roughly equivalent to our upper thigh? Neither did I. But the humped back and large, pale-colored coxae on my insect are what had looked like a monkey-baby mite to my unaided eye. Fungus gnats are also said to have spinose legs, which must be the scientific way to describe bayonets sticking out of the knees.

I was a little disappointed that none of my entomologist friends could identify my little guy to species. As it turns out, this would require close study of wing venation (okay, that's not too hard), chaetotaxy (which means the arrangement of the bristles on their body), and genitalia (which strikes me as a little too invasive for such a recent acquaintance). Plus, with more than 3,000 species in Mycetophilidae, it might take weeks to get through a dichotomous identification key.

"Most of their natural history secrets remain untold," wrote Peter H. Kerr in his entry on fungus gnats for the Springer.com *Encyclopedia of Entomology*. That may be so, but scientists know more than nothing. For instance, fungus gnats occur on all continents except Antarctica. Most (but not all) types of fungus gnats feast on the fruiting bodies, mycelia, and spores of fungi. They prefer damp habitats where their favorite fungi grow and sometimes form thick swarms. In those forests they play an important role in the food web.

A few species of fungus gnats become pests in the damp soil of gardens, farms, nurseries, and overwatered flower pots. Most of the time, though, a female fly will lay her eggs—up to 1,000 during her weeklong life—in the cap of a freshly sprouted mushroom. The larvae develop quickly (three weeks from egg to adult) while burrowing into the cap or making sticky webs on its underside. A few types of larvae are semipredacious and may eat other insects who visit the mushroom.

Later I took my mushroom guy's advice about Googling "glowworms." Radiant turquoise jewels dripping from cave ceilings appeared in Google Images. As it turns out, in a related family of fungus gnats about a dozen species have developed bioluminescence. They mostly live in sheltered grottos in New Zealand and Australia. There, tiny larvae spin nests out of silk on the ceiling and dangle dozens of threads of silk beaded with droplets of mucus. Breathless, calm habitats are necessary so that their lines don't get tangled. Breathless, I'm sure, are explorers who find a replica night sky illuminated on the ceiling of a cave.

The gnat larvae's glow results from luciferin, a chemical compound similar to that used by fireflies. A hungry larva glows brighter than one who has recently eaten, and that glow lures midges, mayflies, mosquitoes, and moths to their doom.

A trip to New Zealand may be in order someday, but for now I'm just happy to know that a pack of mosquitoes wasn't still trying to invade my house this fall!

Five Things That Improve After a Hard Freeze

Vicky Zalatoris

1. Sparkling rubies nestle in emerald green moss, illuminated by the weak glow of a headlamp. Dampness oozes from the deep layers of peat soil and from the fresh layers of sphagnum that climb the scaffolding of twiggy leatherleaf bushes. And my boots. The dampness also oozes around my boots.

Reaching, stretching, I search almost manically for the little bog jewels. Ignoring backaches, ignoring the way that damp moss and damp breeze suck the warmth from fingertips, ignoring a friend who is not ignoring those things and wants to go home. The one-quart yogurt tub in my left hand has almost been filled with cranberries by my right hand. Almost. Not quite, and look, there's another patch by that hummock. And another over there. That one is pale and firm—not yet frosted and sweetened. Look! Three perfect jewels dangle among dry threads of grass.

The first wet snowstorm of fall had pushed marble-sized berries deep into the moss. Then the snow melted. More is forecast, so we must pick everything we want now under the cloudy darkness, listening to the rush of city traffic not far beyond the soggy bog and tangled forest.

A cherry red gem gleams from its cozy bed. I pluck it with numb fingers, listen to it plunk onto its comrades. In its hole I spy more red. I scoop that one up too, only to see a third fruit shining from even deeper. This one is so perfectly ripe that I place it on my tongue, gently pressing it against the roof of my mouth until it bursts. No one sees my pucker face in the dark. Nor do they hear the involuntary hum of satisfaction as the frost-softened, cold-sweetened, pungently wild flavor of the bog fills my night.

2. The hard edge of my library card peels elegant curls of frost off the windshield. A quick peek under the garden cold frame reveals a small grove of deep green spinach—frozen solid. Oh well; next year I'll eat them sooner. But hard frosts require a clear sky, and the day soon warms. Later I find my spinach cheerful, thawed, sprightly, and intact. Into the salad bowl quick, before it's too late! Unlike the wilted and bitter salads of summer, these leaves are sturdy and sweet, requiring much grateful mastication in my sunny kitchen.

3. A friend tells me that his delivery guy drove through a sideways storm. It's gone now, but the northwest sides of the tree trunks are white—skunk striped with damp, driven flakes. Our hiking boots make sculpted designs in the snow. The woods are silent and still. We see only red squirrels and one hairy woodpecker. But others have been here recently. Otters, fishers, foxes, voles, and shrews crossed our path. The snow holds the record of their presence and their passing. Otherwise how would I know? Snow makes the forest feel alive.

4. In the sunny opening created by a fallen sugar maple, tan moths flutter feebly. Moths? But it can't be more than 40 degrees. Quick! Get a photo! Here, I caught one. Delicate fringes outline their wings while darker brown fuzzy and scalloped lines fill them in. They are everywhere. On the snow. Deeper in the woods. Drifting in front of our noses.

Bruce spanworms are pesky little green and yellow inchworms who turn unfurling spring leaves into lacy green skeletons. They prefer sugar maple, aspen, elm, and apple. In June they pupate and wait until cold drives away hungry bats, birds, and predacious insects. In the relative safety of the late fall woods the adults emerge. Wingless females, full of eggs instead of flight muscles, simply crawl up the nearest tree and waft pheromones into the breeze. They wait. Meanwhile the males give thanks for antifreeze and shiver large flight muscles up to temperature. The cold dictates relatively slow wing beats, but those wings are broad enough to carry a lightweight body through the air. Females lay pale green eggs into the grooves of bark. Neither adult eats. Over winter the eggs turn orange. Unfazed by the cold, caterpillars hatch in May.

5. Mosquitoes. My deepest gratitude to whichever source of infinite wisdom reserved the power of winter flight for soft-winged moths and not mosquitoes. Praise be to the hard freeze. Praise be to the delicacy of their little, buzzing lives.

Wood Wide Web I

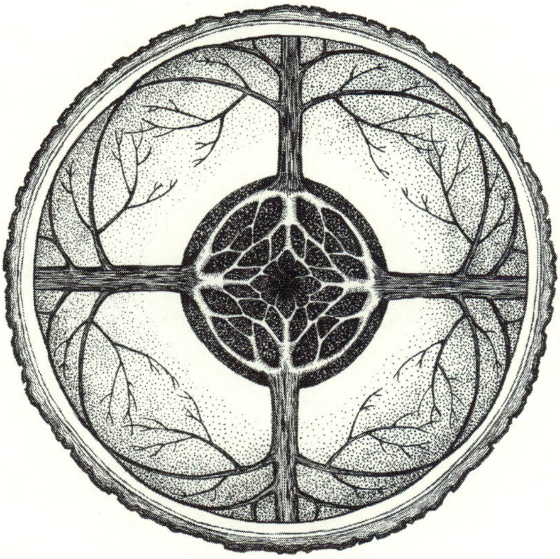

Janet Moore

The forest is quiet and still. Overnight the first snow transformed the woods from a sepia-toned landscape to a black-and-white photo. Looking up, I see how the lacy pattern of twigs embosses a network across the sky. When an icy shard catches me in the eye, I look back down and am struck by the contrast. At the level of the forest floor each trunk stands stoically alone, separated from its neighbors by a barren expanse of white.

The reality is quite the opposite. At the level of twigs, trees are separate. To avoid competition for sunlight, they tend to grow away from one another, each seeking their own personal bubble. Occasionally two trunks or two branches will miscalculate and intersect to squeak in the wind or groan as scar tissue melds them together until death do they part. The real connections, though, happen beneath the duff—almost beneath our notice.

Unlike the twigs, who studiously pretend to ignore each other unless forced to mingle, the miles of roots intersect regularly underground in their search for water and nutrients. When roots of the same or closely related species meet under the right conditions, they fuse together. This grafting requires that tree roots be sufficiently large, be in prolonged contact with each other under pressure that crushes their "skin," and have sufficient moisture. Once the subterranean stars align, the grafted roots can transfer sugars, water, minerals, and chemical signals from tree to tree. According to German forester Peter Wohlleben, in his book *The Hidden Life of Trees*, "Most individual trees of the same species growing in the same stand are connected to each other through their root systems."[3]

This connectivity creates stability for the forest at many levels. On a physical level, the interwoven tangle of roots provides stability in windstorms. Imagine yourself alone, feet tight together, being pushed. Either you fall over, or your "trunk" splits and you step forward for stability. If you stand with a wide base or link limbs with friends, you can withstand more force. As trees toss in a gale, their own spreading roots provide a stable base. But it doesn't stop there. Their roots are connected to their neighbors' roots, and their collective base is as wide as the entire forest.

That entire forest is important. Together all the trees

create a microclimate: their very own ecosystem that buffers wind, stays cooler in the summer, and holds moisture at multiple levels. This microclimate is so important that it might even be worthwhile for well-located trees—those with ample access to water, nutrients, and sunlight—to redistribute those resources among weaker trees and prolong their lives. The partnership would serve to maintain stand integrity against pests, disease, and disruption of the microclimate. It might also maximize use of resources by allowing all trees to grow at an ideal rate, instead of having a few wealthy trees hoard resources they can't immediately use while weaker trees slow their growth in response to scarcity. The network of grafted roots facilitates that sharing.

This cooperative community of the grafted root network might confer several additional benefits on its members. For one, it is mostly trees of the same (or sometimes closely related) species that graft at their roots. This means that when stronger trees help weaker ones, they are still focusing on the survival of their own species. The resulting healthy neighbors may become a source of genetic diversity when pollen flies in spring. In addition, by helping weaker neighbors, the stronger trees maintain an "enemy" they know and prevent the establishment of an unknown competitor who might eventually reduce their access to sunlight or water.

Finally, when a tree dies its roots can persist and continue contributing to the community. This is no small inheritance in a tree's constant search for resources.

Although the orphaned roots can't deliver sugars from their own crown anymore, they can bring up nutrients and water in exchange for maintenance by the community. We can see the result of this exchange in the existence of living stumps. Although not every stump will continue to grow, I have occasionally observed new scar tissue forming on the cut surface. This is evidence that while its crown is gone, the cut tree's root system is still well connected and active.

In an even more remarkable example of network support, small redwood, spruce, and balsam fir trees can be albino. Their needles are ghostly pale and contain no chlorophyll. All their sugar, then, must come from other trees, intravenously fed through their roots.

This period between Thanksgiving and the New Year—when many of us open our checkbooks to share resources with those in our community who have less or donate to organizations that we feel will support our favored microclimate—has many parallels to the forest. While we may sometimes feel like stoic individuals, an interwoven community is hidden just out of sight.

Wood Wide Web II

Margo Butler, age 16

Winter is coming in fits and starts. Last week a thick, white blanket of snow separated each stoic tree trunk. Then it melted. This morning a thin frosting of damp flakes swirled out of the sky, outlining every feature of the forest floor, highlighting the trunks and tracing the twigs. The intricate patterns thus revealed give us a glimpse of the complexity that lies beneath. Yes, the roots of the trees ramble widely and fuse with relatives to form a "wood wide web" that shares resources, but the connections don't stop there.

Fungi pick up where the roots leave off. Most tiny root

tips are colonized by filaments of fungi, which extend the plants' reach into the surrounding soil. Each tree species can host multiple types of fungi, and each fungus can colonize a variety of plant species. This expands the network beyond the species-specific root grafting and creates a mycorrhizal (myco = fungus; riza = root) network that is robust and thickly woven. So thickly, in fact, that scientists say you could find seven miles of fungal hyphae in a pinch of dirt and hundreds of miles under a single footstep.

When Dr. Suzanne Simard (a forest ecologist from the University of British Columbia who coined the term "wood wide web") used DNA to identify all the individual trees and fungal strains in a forest plot, she ended up with a diagram resembling the airline route map in your seat-back pocket.[4] A big, old, well-connected white pine is like Chicago O'Hare, because it provides connections to lots of other places. Hub trees connect (through the mycorrhizal network) to other hub trees, and also to younger, more isolated trees. This pattern is incredibly efficient. The two most distant trees in Simard's study plot needed only two "flights" to connect, and the maximum path length the researchers discovered required only three "flights."

This pattern creates resilience. With just a few important hub trees there is a lower likelihood that one of them will be damaged. If one of the less connected trees dies, the impact on the ecosystem isn't large. However, in the infrequent event that one of the few hub trees dies, the impact on the connectivity of the surrounding forest could be dire. That's why conservationists and foresters must focus on identifying and preserving the venerable elders.

Trees and mycorrhizal fungi live in an incredible symbiosis. Trees feed the fungi sugars produced during photosynthesis and may share up to 80 percent of their total production. In return, fungal hyphae increase the absorbing area of roots from 10 to 1,000 times. This protects trees against drought by contributing to water uptake and storage. Fungi don't just absorb nutrients, they actively break down tightly bound soil nutrients like phosphorous and iron and make them available to plants. Some fungi even recycle nitrogen from soil nematodes into their plant hosts. Fungi and trees feed each other, and the intricate web captures and holds nutrients before they can be lost from the system.

The benefits aren't just food related. The sheath that a fungus forms around a root creates a physical barrier against diseases. Meanwhile, it is diverting heavy metals and excreting antibiotics designed to kill pathogens. At a community level the fungi act like fiber-optic cables. Using warning chemicals and electrical impulses, they help trees share information about insects, drought, and other threats so that currently unaffected trees can prepare their defenses in advance of an invasion.

What's more, the mycorrhizal fungi can facilitate nutrient sharing between trees and plants of different species. Simard traced the flow of radioactive carbon 14 from birch to Douglas fir trees. After only an hour she found evidence that birches had incorporated the carbon 14 into sugars during photosynthesis and then shared it with nearby firs (which had been shaded for the experiment). Later, after leaf-off, she found that the evergreen firs were providing sugars to the twiggy birches.

Why would the fungi benefit from distributing resources to trees in need? Their existence depends on the presence of a mature, stable forest with its humid microclimate and flow of nutrients. Tree diversity is essential, since monocultures are vulnerable to disease and disturbance. Therefore, the fungi may share resources in an effort to maintain diversity in the forest and ensure the long-term stability of the environment they enjoy. A diversity of fungi is also important, since different fungi provide different benefits to their hosts and thrive in different situations.

The wood wide web, with all of its natural connections, is beautiful from many angles. Perhaps human society could do worse than try to emulate a system that has kept forests thriving on Earth for millennia.

The Elegant Simplicity of Moss

Michelle Allison

Slushy snow gave way to bare, squishy gravel as I crested the hill. With the surface change, the rubber and metal grippers on my running shoes instantly became annoying instead of essential. Peering wistfully through the damp, brown woods, I could just make out the wide clearing of the Birkie ski trail. But my skis were still in the basement, waiting for real snow.

Brown, brown, brown. The forest appeared drab and dead. But as I looked up from the effort of the hill climb, vibrant greens glowed into view. A scattering of small boulders, probably dumped here by the glaciers and later excavated when the road was built, hosted emerald carpets of life. Then, squish. One moment of distraction and my foot found a puddle. I thought the season for puddles had passed.

Of course, even though this is the warmest fall on record across the globe, I can't directly blame today's warm weather on climate change. Climate is what you expect. Weather is what you get. They don't often match up perfectly in day-to-day life.

That's why we have to be ready for anything. Moss is. It's thriving today in the cold mist. The persistent evergreen leaves of mosses are able to take advantage of favorable growing conditions in any season. Even when drought withdraws the water they need for growth, mosses are preparing for life in the future. Essential functions shut down and prepare for dormancy. Cell membranes shrink like a vacuum-sealed freezer bag. And, with amazing "forethought," the mosses synthesize and store away the enzymes of cell repair that will manage the damage of desiccation. Like the Red Cross or the Federal Emergency Management Agency (FEMA), mosses like to have a stash of medical supplies ready to go. All this groundwork pays off. In just 20 minutes bone-dry moss can return to full vigor. This resilience of mosses is mostly due to their amazing ability to live thriftily and within their means.

Robin Wall Kimmerer, a botanist and author of the award-winning book *Braiding Sweetgrass*, wrote an earlier book called *Gathering Moss*. In this ballad of love to the mosses, she writes, "They are the most simple of plants, and in their simplicity, elegant."[5]

Elegant indeed. Resourceful, one-cell-thick leaves allow water to soak in directly to where it's required without the need for constructing expensive distribution systems. Moss doesn't even have roots. They don't need to suck resources out of the ground. Their tiny rhizoids only serve to anchor them to the substrate. Being filled with water is not enough. Moss also needs a film of rain—or melting snow—to cover the outside of the leaf and to act as a conduit for carbon dioxide to enter the leaf from the air.

"Like a jealous lover," writes Robin, "the moss has ways to heighten the attachments of water to itself and invites it to linger, just a little longer."[6] Living in tightly packed clumps improves moss's water-holding efficiency. So does arranging their branches and leaves so that each space is the perfect size to trap a water droplet using capillary action. Leaf surfaces are textured or pleated or sculpted into hills and valleys to grab water. "This elegant design is a paragon of minimalism, enlisting the fundamental forces of nature, rather than trying to overcome them," Robin observes.[7]

Part of the mosses' minimalism is in their size. By staying small, mosses take advantage of the microclimate inside the boundary layer, which "like a floating greenhouse hovering just above the rock surface" traps water vapor, heat, and carbon dioxide. On a sunny winter day when the air is appropriately below freezing, the boundary layer often provides moss with liquid water. Waste gasses emitted from bacteria and fungi on rotting logs can increase the carbon dioxide in the boundary layer to 10 times the amount in the ambient atmosphere. Thus, moss ensures that it has access to a steady supply of raw materials for photosynthesis.

Mosses are confined to this boundary layer. They thrive within it and cannot survive beyond it. In similar fashion humans are restricted to a thin zone of habitable conditions that surround the Earth. We aren't as good as the mosses at living within our means, though.

Robin, with her Potawatomi heritage, talks about stories from the oldest days "when all beings shared a common language." Not anymore. That language is forgotten. Instead, she says, "we must learn each other's stories by looking, by watching each other's way of living. [Mosses] have messages of consequence that need to be heard."[8] The big question is, when will we finally listen?

Camp Robbers and Broken Refrigerators

Owen Erickson, age 13

The thick carpet of pine needles felt springy and dry under the trees. Patches of sunlight kindled warm fragrances to tickle my nose. On the wooden wildlife observation deck, though, an inch or so of soggy snow—not fit for November—dripped steadily through the planks. Skims of ice clung only to the shallowest pools in the marsh, and in one bay a column of mist rose from the dark water. Although it was barely past noon, the low angle of the sun illuminated the swirling cloud in soft light as it danced in partnership with the breeze.

A raucous "kreh kreh kreh" erupted from the edge of the pines. My first thought was a merlin—a small falcon with a big voice. But merlins head south in September, and the peak of their migration has long passed. I peered into the pine boughs where the sound originated and caught a glimpse of a gray-colored bird a little bigger than a robin. As soon as I spotted it the bird flew. My visitor didn't hightail it into the bushes, though; it made an unhurried swoop toward me on downward angled wings and perched in a nearby birch. There was no mistaking that flight pattern; the alarmist was a gray jay.

The habitat range of these nonmigratory birds—once called Canada jays—encompasses most of Canada up to tree line and scoots just under the nose of Lake Superior like a mustache. Here at the southern edge of their range they aren't very common. Usually I have my annual encounter with a "camp robber" in the Boundary Waters Canoe Area Wilderness or Sax-Zim Bog north of Duluth, Minnesota.

Why do they live so far north? Gray jays seem to depend on two things: a strong presence of certain tree species (black and white spruces and jack pine in northern Wisconsin) and cold temperatures. Both requirements relate to their food storage method.

To survive the long, bitterly cold winters in their range, gray jays must put up food all summer and fall. They accomplish this by coating mouthfuls of food in sticky saliva, then gluing the boluses in tree crevices, under lichens, in evergreen needles, and behind the flaky bark of their preferred tree species. A gray jay may hide 1,000 separate caches of food in a single 17-hour day. Scatter-hoarding is the technical term for this technique. "Gray jays have a memory like a Vegas card counter," wrote Joe Rankin in *Northern Woodlands* magazine, referring to the jays' ability to retrieve 80 percent of their food-filled saliva balls.[9]

At least a portion of the lost 20 percent is due to spoilage. Some boreal tree species may contribute antibacterial compounds that help the food stored under their bark stay fresh. But that's not enough. Cold temperatures in the gray jay's preferred habitats are necessary to prevent cached food—including chunks of meat—from spoiling.

On this lovely, sunny, and warm November day the vulnerability of these southernmost gray jays was driven home. Their refrigerator is broken. Each autumn seems to be warmer than the last, and that doesn't bode well for the longevity of cached meat and other food. Their reproductive rates may suffer, since gray jays nest in late winter and feed nestlings from their caches. A study by the University of Guelph in Ontario confirmed that gray jays are food-limited during the breeding season. The researchers connected this with warmer fall temperatures allowing food to spoil and a 50 percent decline in numbers of gray jays in Algonquin Provincial Park over the past three decades.[10]

The decline is not because they're picky eaters either. Gray jays are opportunistic omnivores, and they consume everything from small mammals, nestling birds, carrion, and arthropods to fungi, fruits, and seeds. They have been observed picking engorged winter ticks off the backs of moose. Anyone who has camped within their range knows that these camp robbers are not shy about snatching a crust of bread off the picnic table or nabbing a marshmallow before it can become a s'more.

My lack of food made me boring to the camp robber who came to investigate. He scrutinized me for only a minute before flying off. Since gray jays mate for life and stick together year-round, I would guess that this solo guy is a bachelor, and probably a young of the year. Even without the broken refrigerator, his is a tough lot.

In early June the two-month-old chicks begin an intense sibling rivalry. Only one dominant juvenile gets to stay with its parents after driving the rest of the young away. These "stayers" gets access to their parents' food caches and mentorship in exchange for helping to raise little brothers and sisters the next summer. The "leavers" who get kicked out may be able to volunteer their future nanny services to an unrelated pair who is currently without chicks of their own. Or they may simply perish during the hungry winter. Leavers experience an astounding 85 percent mortality rate, while stayers die only about half the time.

As I gazed over the wetlands and pine forests the midday sun warmed my face and the unseasonable weather made me sigh. That lonely jay has every right to scold us for breaking his refrigerator.

Where Are the Loons?

Susan Lewis

Temperatures plummeted beneath a black dome of sparking stars. As the eight inches of ice already on the lake thickened and expanded in the cold, it sang the eerie notes of winter. A couple of weeks ago the ice notes were high and squeaky—shorter frequencies due to thinner ice—but tonight they deepened into the lower moans of thick ice. The ice's plaintive wails brought to mind the lake songs of summer—as performed by loons.

All summer long their human neighbors alternately curse and acclaim their calls. One tired and grumpy visitor asked me, "Do loons ever sleep?" after nearby birds had caterwauled into the wee hours of the morning. Although they do sleep, it's not on our schedule. Since loons are visual hunters, they must feed when sunlight penetrates into fish habitat. When darkness brings relative quiet to the woods, they use that opportunity to communicate with neighbors.

"Who's there? I'm here" may be the basic message in their wail. By speaking up, each loon reasserts that it is on its territory and ready to defend against invaders. If a loon remains quiet, that could signal that its territory is free for the taking, a risky message to convey. As male loons patrol their territorial boundaries each night, encountering a male intruder will trigger a yodeling contest. That's the maniacal cry, called "the laugh of the deeply insane" by writer John McPhee, that you might curse at three a.m. It may or may not result in a physical battle.

Where are the loons in late November, though? I haven't heard them calling since August or seen an adult in quite a while. The last loon I saw was the sole surviving juvenile from a nest in a nearby bay. With any luck, he was able to gain enough strength to migrate south before the lake began to freeze.

There's no way of knowing where my particular unmarked loon neighbor went, but the US Geological Survey has captured and radio marked several juvenile loons on lakes scattered across Minnesota and Wisconsin. Researchers use geolocators and satellite transmitters to track the movements of migrating and wintering loons.

Juveniles tend to begin their journeys in late October through mid-November. Their routes aren't identical, and some juvenile loons take layovers on Lake Michigan.

Others make pit stops on the Mississippi River or reservoirs in the central states. The juveniles must travel alone, since their parents and other nonbreeding adults migrate earlier. In general, nonbreeding adults start migration in late August, mothers leave in September, and fathers head out in October.

The ultimate destination for many Wisconsin and Minnesota loons is the Gulf of Mexico, and specifically the west coast of Florida. Their main habitat requirements are plenty of fish to eat and clear water in which to hunt. Southern inland lakes tend to have warm, shallow, murky water, and alligators, so the ocean provides a better option. There loons face the challenge of transitioning from freshwater to saltwater. They've adapted by excreting salt out of glands in their skull between their eyes. The glands drip almost constantly during the winter . . . sort of like how my nose adapts to winter too . . .

By making the big journey and coping with salt, loons gain access to a seafood feast. Wintering loons eat flounder, crabs, lobster, shrimp, gulf menhaden, bay anchovies, silversides, and more. The ocean bounty gives loons enough energy to molt and regrow all their feathers, but danger is still present. Not much is known about loon predation in the ocean, but I've heard that dog sharks will race upward through the water to attack loons from below.

Unfortunately for human snowbirds, the loons' winter feast doesn't happen within viewing distance from beaches. Most loons spend the winter at least five miles offshore. Even if you could see the loons, you might not recognize them. The adults and juveniles all wear drab brown plumage and sport dark eyes instead of the ruby-red eyes of breeding season. They aren't known to vocalize. In the spring adults molt back into breeding plumage, complete with the red eyes. Juveniles stay brown and ocean bound while they mature and will migrate back toward their natal lake only after they are three to four years old.

Loons need to spend winter on the ocean, but I am grateful that human adaptations allow me to enjoy my northern lakes all year round. When the haunting calls of the loons fade with the summer sunshine, the lake itself sings an icy winter tune. Perhaps it too is lamenting, "Where are the loons?"

Willow Pinecone Galls

Susan Lewis

Snow crunched under my tires and bare trees whizzed by as I coasted down the hill on my fat bike. Cold air stung my cheeks, but my hands were warm in homemade pogies that wrapped around my hands and bike handlebars like old-fashioned muffs. At the bottom of the hill I passed through a wetland with dried grasses and shrubs frozen into the ice.

In the slightly drier ground next to the graveled forest road grew sparsely elegant clumps of willows, including an unusual one that looked like it had hybridized with a pine tree. At the tip of most of the twigs perched a compact little "pinecone," about the size and shape of a young, tightly closed cone from a red pine. These cones don't hold any tree seeds, however. Instead, they harbor the "seeds" of a gall midge called *Rhabdophaga strobiloides*. These are willow pinecone galls.

Like most galls, these began in early spring during the active growth period for the plant and the egg-laying season for the midge. The adult midge laid an egg on the tip of the twig, right where a single bud scale protected baby leaves. The larva hatched in early May and started burrowing into the willow stem. Some combination of the chewing action and saliva of the newly hatched larva triggered an increase in plant growth hormones. Cells grew bigger and more plentiful, but the stem did not extend. Instead, leaves once destined to flutter along a twig now layered together in the cone-like structures that caught my eye.

What's fascinating is just how much the larva can control the plant. One study found that the twigs hosting a gall were larger in diameter than twigs with no gall—even if the twig did not have leaves. Bigger stems were correlated with bigger galls, and bigger galls were correlated with bigger larvae. This confirms the hypothesis that the larva somehow draws in the products of photosynthesis from other ungalled twigs to spur the growth of the gall, the hosting twig, and the larva itself.

This type of control may seem creepy, but it is common in the world of parasites. Parasitic *Cordyceps* fungi force ants to climb a plant and attach there before they die, providing a breezy platform for dispersal of the fungal spores from the ant's corpse. A brain parasite causes rats to be attracted to cats so that the parasite can complete its life cycle in a feline host. Horsehair worms drive zombie crickets to a watery death. In comparison, a tiny larva causing a plant to grow some extra tissue is pretty tame.

It is still amazing. All summer the larva—eating stolen nutrients from the plant—grew within the protective walls of the gall. Late last May the larva would have expanded enough that it needed to shed its skin, thus entering its second instar developmental stage. By late July the growing larva shed its skin once more. Its next task before winter was to construct a cocoon and leave it open at the top.

Inside each of these cone-like galls by the side of the road is a little larva in a cocoon sleeping bag, steeling itself against the cold. The loosely packed structure of the gall may provide some insulation, but it isn't nearly as good as my homemade pogies. The dormant larva doesn't have the benefit of my fiery metabolism fueled by regular eating to keep it warm. In these situations, some creatures allow themselves to freeze. Wood frogs are one incredible example.

These larvae take a more daring route. By concentrating glycerol (once used in cars as antifreeze) in their bodies, the larvae can supercool their liquids down below the freezing point of water without them becoming solid. In extreme examples, larvae have safely supercooled to negative 76 degrees. This requires a delicate balance. A disturbance in the system could cause the larvae to freeze instantly and die.

If a larva survives the winter, it will pupate inside the preformed cocoon early in April, and the adult gall midge will squeeze out between the layers of the cone and fly away.

Or not. Thirty-one other creatures might be illicit squatters in the loosely packed scales of the pinecone gall. These gall apartments are in such demand that during one study the naturalists found 564 individual insects in just 23 galls.[11] Most of the beetles, caterpillars, sawflies, and eggs of meadow grasshoppers are relatively harmless. The wasps, however, beat the gall midges at their parasitic game by raising their wasp larvae on the tender flesh of midge larvae.

Who would have thought that the scales on those strange "pinecones" could be hiding such drama?

Little Kings of the Winter Woods

Kaila Core, age 8

A light wind rattled the oak leaves still clinging to their twigs while my snowshoes crunched through the crust, and expanding lake ice sang eerily in the distance. So far my walk—a quest for inspiration—had yielded only cold cheeks and a stark winter woods.

Near the end of the trail a new sound filtered gradually into my consciousness. High, thin "see see" notes drifted down from the dense foliage of a little stand of fir trees. Chickadees I thought at first, and smiled, although I did not look up. Then I caught a movement out of the corner of my eye. Just above my head several little birds fluttered in the boughs. They seemed too delicate to be chickadees, and their song was too high and thin, so now I focused in on their field marks.

Instead of the black chin of a chickadee, I looked up at a cream-colored throat and breast. Their backs and wings were warm olive with black and yellow markings. I started to develop a hunch about who might be in this little flock, but it took me several minutes of focused observation (of course I'd forgotten to carry binoculars) before one tipped her little head right toward me. Aha! A yellow stripe ran down the top of her head, black stripes bordered it, with another black stripe through the eyes. Golden-crowned kinglet.

Now here was some inspiration. Golden-crowned kinglets are one of the smallest songbirds, and yet they commonly overwinter in places where the nighttime lows plummet to 40 degrees below zero. Bernd Heinrich, a professor of mine from the University of Vermont, wrote about them extensively in his book *Winter World*. He mused, "Only if I knew how and why a golden-crowned kinglet survives a Maine or Alaskan [or Wisconsin!] winter would I understand the story of winter survival. . . . The kinglet is thus iconic not only of winter, but also of adaptability under adverse conditions."[12]

I inhaled deeply, a breath of gratitude and wonder at coming across these little miracles of adaptation, but the dry, 11-degree air caught painfully in my lungs. The kinglets showed no such distress despite their acrobatic and aerobic exercise. Not only are their bird lungs quite different from mine, each of their delicate nostrils is covered by a single tiny feather. I can only lament that I didn't wear a scarf.

Crunching around to get a better look at the flock, my snowshoes caught awkwardly on twigs and crust. Help as they do, snowshoes can't quite float my large mammal body above the drifts. From my new vantage, though, I got a great look at one of the flock fluttering in midair. He could hover under the twig to feed, just like a hummingbird. Then with an effortless swoop, he landed on the tip of a different twig, his two-penny weight barely bending the tree as he dipped his head in continuous feeding motions.

Kinglets need to feed constantly, at almost one peck per second, discovered Bernd. Why this apparent gluttony? To maintain their internal furnace at 110 degrees, kinglets need to eat two to three times their own bodyweight of energy-dense food each short winter day.

What could kinglets possibly be finding to eat in this barren landscape? Bernd wondered that too. His research included dissecting a kinglet to examine the gizzard contents—inchworms—then examining the trees themselves to find living examples of those same inchworms, and finally raising the inchworms to adulthood so that they could be identified. The answer: kinglets often eat caterpillars of the one-spotted variant moth (*Hypagyrtis unipunctata*). These happen to be relatives of the Bruce spanworm moths I encountered recently, and I can't help but wonder if the spanworm moths or their eggs might also be food for winter kinglets.

The slight breeze cut through even my puffy winter

coat. How are those tiny birds staying warm? Kinglets don't have a big enough body to store fat like a bear does, but they do grow more feathers each autumn. During the winter, feathers make up 8 percent of their body weight. Artic explorers wear about the same relative weight of clothing. Like my jacket, those feathers are great insulation.

But their feet aren't feathered, I noted, wiggling my warm toes inside wool socks and thick boots. And that is where any desire I might have to be a kinglet stops. Their feet are always cold. It isn't worth the effort to keep them warm, as long as they aren't frozen and damaged. Their main worry is that cold blood returning to the body from the feet might lower their core temperature. The solution is a countercurrent heat exchange system where arteries with warm blood from the body passively warm up blood from the feet before it reenters the body.

This evening the kinglets will continue with their constant search for food. They'll feed until the last instant before dark, then crowd into whatever shelter they can find, and huddle up with the flock. Those are their instructions for life. My instructions? Seek stories in the winter woods, but make sure you have warm feet.

Notes

WINTER

Hairy Woodpecker in the Fog

1. ZhaoDan Zhu, Wei Zhang, and ChengWei Wu, "Energy Conversion in Woodpecker on Successive Peckings and Its Role on Anti-shock Protection of Brain," *Science China Technological Sciences* 57, no. 7 (July 2014): 1269–75, https://doi.org/10.1007/s11431-014-5582-5.

Woodpecker Niches

2. Nell Durfee, "There's a New Theory for Why Downy and Hairy Woodpeckers Look So Alike," *Audubon*, February 26, 2018, http://www.audubon.org/news/theres-new-theory-why-downy-and-hairy-woodpeckers-look-so-alike.

Grouse House

3. Guthrie S. Zimmerman, Rick R. Horton, Daniel R. Dessecker, and R. J. Gutiérrez, "New Insight to Old Hypotheses: Ruffed Grouse Population Cycles," *Wilson Journal of Ornithology* 120, no. 2 (2008): 239–47.

Surrender

4. David George Haskell, *The Forest Unseen: A Year's Watch in Nature* (New York: Penguin Books, 2013), 2.
5. Ibid.

Winter Houseguests

6. Larry Weber, *Spiders of the Northwoods* (Duluth: Kollath-Stensaas, 2013).
7. Wikipedia, s.v. "Tegenaria," accessed April 14, 2018, https://en.wikipedia.org/wiki/Tegenaria.

8. "Giant House Spider (Tegenaria gigantea)," Woodland Park Zoo, accessed August, 20 2013, https://www.zoo.org/document.doc?id=203.

9. David Sedaris, *When You Are Engulfed in Flames* (New York: Little, Brown, 2009).

Soft, Gray Wings

10. "The Secrets of Owls' Near Noiseless Wings," American Physical Society, last modified November 24, 2013, https://www.eurekalert.org/pub_releases/2013-11/aiop-tso111813.php.

Goldenrod Gall Moth

11. William E. Miller, "The Goldenrod Gall Moth Gnorimoschema Gallaesolidaginis (Riley) and Its Parasites in Ohio (Lepidoptera: Gelechiidae)," *Ohio Journal of Science* 63, no. 2 (March 1963), http://hdl.handle.net/1811/4921.

12. Ibid.

13. Ibid.

Wolves at Rock Lake

14. Adrian Wydeven, email message to author, 2015.

15. Ibid.

SPRING

A Prickly Botanist

1. Uldis Roze, *The North American Porcupine* (Ithaca, NY: Cornell University Press, 2009).

The Pulse of a Waking Forest

2. H. E. Tuttle, "Drumming of the Ruffed Grouse," *The Auk* 26 (1919): 337.

3. "Ruffed Grouse Overview," Cornell Lab of Ornithology, accessed April 15, 2018, https://www.allaboutbirds.org/guide/Ruffed_Grouse/overview.

Northwoods Monkeys

4. Arthur Cleveland Bent, *Life Histories of North American Woodpeckers*, Bulletin 174 (Taunton, MA: Smithsonian Institution United States National Museum).

5. Ibid.

The Changing Woods

6. Scott R. Loss, "Relations among Invasions of Non-native Earthworms, Forest Floor Habitat, and Populations of Ground-Nesting Songbirds in North Temperate Hardwood Forests," (PhD diss., University of Minnesota, 2011), https://conservancy.umn.edu/handle/113203.

7. John C. Maerz, Victoria A. Nuzzo, and Bernd Blossey, "Declines in Woodland Salamander Abundance Associated with Non-native Earthworm and Plant Invasions," *Conservation Biology* 23, no. 4 (2009): 975–81, http://www.jstor.org/stable/29738833.

Bird Muscles and Migration

8. "Depression," Sparking Life, last modified December 14, 2016, http://www.sparkinglife.org/2016/12/14/depression/.

9. Steve Bouser, "Return of the Wood Thrush's Song," NPR Opinion, last modified May 21, 2010, https://www.npr.org/templates/story/story.php?storyId=126511080.

10. Bernd Heinrich, *Why We Run: A Natural History* (New York: Ecco, 2002).

11. Ibid.

Winter Wren

12. Irby Lovette, "The Winter Wren's Wraparound Range Map," All About Birds, last modified April 15, 2009, https://www.allaboutbirds.org/the-winter-wrens-wraparound-range-map/.

SUMMER

Boys on the River: Imitators and Attractors

1. L. M. Montgomery, *Anne of Green Gables* (New York: Bantam Books, 1976).

Pitcher Plant Partnerships

2. Chel Anderson and Adelheid Fischer, *North Shore* (Minneapolis: University of Minnesota Press, 2015).

The Brookies of Cap Creek

3. Patrick D. Shirey, Jill Medland, Gary A. Lamberti, "Namekagon River Cold Water Zone Fish Habitat History Report," June 2009, https://www.nps.gov/sacn/learn/nature/upload/Fish-Habitat-History-Report-6-24-09.pdf.

Forest Bathing

4. "Can Bacteria Make You Smarter?," American Society for Microbiology, ScienceDaily, accessed April 20, 2018, www.sciencedaily.com/releases/2010/05/100524143416.htm.

Growing Up Before My Eyes

5. Charley Eiseman, Noah Charney, and John Carlson, *Tracks and Signs of Insects and Other Invertebrates* (Mechanicsburg, PA: Stackpole Books, 2010).

The Channels of Bark Bay Slough

6. Adelbert Llewellyn Leathers, *Ecological Study of Aquatic Midges and Some Related Insects with Special Reference to Feeding Habits* (Washington, DC: Government Printing Office, 1922).

Tongues of the Earth

7. L. Margulis and D. Sagan, *What Is Life?* (Berkeley: University of California Press, 2000).

8. Rachel Carson, *The Sense of Wonder* (New York: Harper and Row, 1965).

FALL

All Flourishing Is Mutual

1. Robin Wall Kimmerer, *Braiding Sweetgrass: Indigenous Wisdom, Scientific Knowledge and the Teachings of Plants* (Minneapolis: Milkweed Editions, 2013).

2. Ibid.

Wood Wide Web I

3. Peter Wohlleben, Tim F. Flannery, S. Simard, and Jane Billinghurst, *The Hidden Life of Trees: What They Feel, How They Communicate: Discoveries from a Secret World* (Vancouver, BC: Greystone Books, 2016).

Wood Wide Web II

4. Kevin J. Beiler, Daniel M. Durall, Suzanne W. Simard, Sheri A. Maxwell, Annette M. Kretzer, "Architecture of the Wood-Wide Web: *Rhizopogon* Spp. Genets Link Multiple Douglas-Fir Cohorts," *New Phytologist* 185, no. 2 (January 2010): 543–53.

The Elegant Simplicity of Moss

5. Robin Wall Kimmerer, *Gathering Moss: The Natural and Cultural History of Mosses* (Corvallis: Oregon State University Press, 2004).

6. Ibid.

7. Ibid.

8. Ibid.

Camp Robbers and Broken Refrigerators

9. Joe Rankin, "Gray Jays: Birds with Attitude," *Northern Woodlands*, October 2014.

10. Dan Strickland, Brian Kielstra, D. Ryan Norris, "Experimental Evidence for a Novel Mechanism Driving Variation in Habitat Quality in a Food-Caching Bird," *Oecologia* 167 (2011): 943, https://doi.org/10.1007/s00442-011-2040-y.

Willow Pinecone Galls

11. Donald Stokes and Lillian Stokes, *Stokes Guide to Nature in Winter* (New York: Little, Brown, 1979).

Little Kings of the Winter Woods

12. Bernd Heinrich, *Winter World: The Ingenuity of Animal Survival* (New York: Ecco, 2003).

Zac Stone, age 8

Notes about Illustrations

Illustrations for this book were truly a community effort. Since kids are often the inspiration for Natural Connections articles, the Cable Natural History Museum conducted an art contest for local students to illustrate each of the chapters. The hope was that students would read some chapters, learn something about the plants and animals in those chapters, do a little more research on their own, and then create a black-and-white line drawing based on their research. We were thrilled with the response. To fill in the gaps, the Museum opened up the illustration project to artists of all ages and abilities. Friends, relatives, strangers, amateurs and professionals were all tapped to translate Emily's writing into simple drawings. Many thanks to all who contributed to this project!

Note about Proceeds

Proceeds from this book will support natural history education for youth at the Cable Natural History Museum.

Bradan Gingras, age 11

Acknowledgments

Everything in nature is connected. Nothing is created without the support of a zillion other living and nonliving parts of the Universe. This book is no different. It could not be in existence without the trees in the paper; the sources of energy used to write, print, and transport the final product; and the Earth that sustains us all.

Of course, many humans contributed their talents as well. I won't list them all here, but they know who they are. My parents, Margaret and Larry Stone, deserve much credit for bestowing on me their book- and nature-loving genes, for raising a mud-and-water-daughter, for using lots of red pen and encouragement on my high school papers, for continuing to edit my newspaper articles each week, and for immeasurable support in every way imaginable. My brother and sister-in-law contributed four fantastic children to the family, who provide me with much joy and inspiration. My Aunt Nan's spirit still zooms around on dragonfly wings, infusing my world with her special sort of love.

My esteemed teachers, Mrs. Moser, Mrs. Probert, Mrs. Stott, and Mrs. Gnagy, all made learning the nuts and bolts of English quite enjoyable. Rick Penn, my grizzled, curmudgeonly, English 211 professor at Northland College gave me, for the first time, the delightful assignment of a short essay due every week. He had wonderful taste in books. Professor Alan Brew did as well. Many natural history professors, especially Craig

Prudhomme, ignited my spark for interpreting the natural world. Tom Fitz, Jim Meeker, and Mike Link fleshed out those ideas, as did all the brilliant professors and fellow students in the Field Naturalist and Ecological Planning master's programs at the University of Vermont.

I am also grateful to the Cable Natural History Museum community, led by energetic Museum Director Deb Nelson, for their support. The Museum's founding director, Lois Nestel, started a tradition with her Wayside Wanderings newspaper articles, which Sue (Benson) Thurn continued enthusiastically. I'm honored to receive the torch from those remarkable women. Thanks also to the kids (of all ages) who ask great questions and inspire me with their curiosity.

From within the Museum community, several people—Fred Stratton, Jodi and Henry Kingdon, and Ron and Patty Anderson—have generously supported this project financially as well as with kind words and shared excitement. Everyone who supports the Museum is part of this book. Thanks to all of the artists, young and old, who participated in my crazy idea and helped me illustrate this book with their wonderful and whimsical drawings. I'm so glad to have you involved.

Although many have helped, I assure you that any errors still remaining are my own.

And again, thanks to the natural world who is full of many kinds of magic.